Fracture mechanics of concrete: Material characterization and testing

ENGINEERING APPLICATION OF FRACTURE MECHANICS
Editor-in-Chief: George C. Sih

G.C. Sih and L. Faria (eds.), Fracture mechanics methodology: evaluation of structure components integrity. 1984. ISBN 90-247-2941-6.

E.E. Gdoutos, Problems of mixed mode crack propagation. 1984. ISBN 90-247-3055-4.

A. Carpinteri and A.R. Ingraffea (eds.), Fracture mechanics of concrete: material characterization and testing. 1984. ISBN 90-247-2959-9.

G. Sih and A. DiTommaso (eds.), Fracture mechanics of concrete: structural application and numerical calculation. 1984. ISBN 90-247-2960-2.

Fracture mechanics of concrete: Material characterization and testing

Edited by

A. Carpinteri

University of Bologna,
Bologna, Italy

A.R. Ingraffea

Cornell University
Ithaca, New York, USA

1984 **MARTINUS NIJHOFF PUBLISHERS**
a member of the KLUWER ACADEMIC PUBLISHERS GROUP
THE HAGUE / BOSTON / LANCASTER

Distributors

for the United States and Canada: Kluwer Academic Publishers, 190 Old Derby Street, Hingham, MA 02043, USA
for the UK and Ireland: Kluwer Academic Publishers, MTP Press Limited, Falcon House, Queen Square, Lancaster LA1 1RN, England
for all other countries: Kluwer Academic Publishers Group, Distribution Center, P.O. Box 322, 3300 AH Dordrecht, The Netherlands

Library of Congress Cataloging in Publication Data

Library of Congress Cataloging in Publication Data
Main entry under title:

Fracture mechanics of concrete.

    (Engineering application of fracture mechanics ; v. 3)
    Includes bibliographies and index.
    1. Concrete--Fracture. 2. Fracture mechanics.
I. Carpinteri, A. II. Ingraffea, A.R. III. Series.
TA440.F734 1984   620.1'366   84-1678
ISBN 90-247-2959-9

ISBN 90-247-2959-9 (this volume)

Copyright

PRINTED IN THE NETHERLANDS

Errata

Page   **5, eq. (1.3):** $\sigma_c^{(c)} > \sigma_m^{(c)}$ instead of $\sigma_c^{(c)} > \sigma_m^{(t)}$

       12, Table I: ϵ_f (cm/cm $\times\ 10^{-4}$) instead of ϵ_f (cm/cm $\times\ 10^{-3}$)

       21, 1st line: compare instead of prepare

       25, 8th line: $d = 12.7$ cm instead of $d = 72.7$ cm

       26, 2nd line: ligament $d - a$ instead of ligament $d - c$

       35, 1st line: close instead of closed

       35, Fig. 2.4 caption: $\dot{\epsilon}$ instead of ϵ

       51, 37th line: $\theta < \eta \leqslant \pi/2$ instead of $\theta < \eta \leqslant \Pi$

       56, 7th line: 2.30 instead of 2.31

       58, 17th line: $a = c(l + k)$ instead of $a = c(l - k)$

       61, 3rd line: "... is randomly different ..." instead of "... in randomly different ..."

       68, 16th line: was instead of has

       80, 12th line: of LEFM instead of to LEFM

       87, Table 3.2: content instead of intent

       96, Table 3.4: compaction instead of composition

       97, 5th line: σ_r instead of σ_2

       100, 24th line: test instead of best

       122, 10th line: change instead of charge

       128, Fig. 4.16 caption: "... on the size ..." instead of "... of the size ..."

       130, 7th and 8th lines: "... were analyzed using the theoretical model." instead of "... when analyzed the theoretical model."

       132, 5th line: f_c' instead of F_c'

       138, 18th line: even though instead of even through

       143, 6th line: discussed instead of discusses

       152, 32th line: elastically instead of elasticity

       163, 15th line, and 180, 8th line: complementary instead of complimentary

       165, 11th line, and 180, 26th line: 1/1000 instead of 1/1000°

       172, eq. (6.11b): $\dfrac{\partial w}{\partial y}$ (P) instead of $\dfrac{\partial w}{\partial x}$ (P)

       177, 9th line: macroscopic fringes instead of microscopic fringes

       183, 9th line: out-of-plane instead of out-of-phase

Contents

Contents

Series on engineering application of fracture mechanics

Fracture mechanics technology has received considerable attention in recent years and has advanced to the stage where it can be employed in engineering design to prevent against the brittle fracture of high-strength materials and highly constrained structures. While research continued in an attempt to extend the basic concept to the lower strength and higher toughness materials, the technology advanced rapidly to establish material specifications, design rules, quality control and inspection standards, code requirements, and regulations for safe operation. Among these are the fracture toughness testing procedures of the American Society of Testing Materials (ASTM), the American Society of Mechanical Engineers (ASME) Boiler and Pressure Vessel Codes for the design of nuclear reactor components, etc. Step-by-step fracture detection and prevention procedures are also being developed by the industry, government and university to guide and regulate the design of engineering products. This involves the interaction of individuals from the different sectors of the society that often presents a problem in communication. The transfer of new research findings to the users is now becoming a slow, tedious and costly process.

One of the practical objectives of this series on *Engineering Application of Fracture Mechanics* is to provide a vehicle for presenting the experience of real situations by those who have been involved in applying the basic knowledge of fracture mechanics in practice. It is time that the subject should be presented in a systematic way to the practicing engineers as well as to the students in universities at least to all those who are likely to bear a responsibility for safe and economic design. Even though the current theory of linear elastic fracture mechanics (LEFM) is limited to brittle fracture behavior, it has already provided a remarkable improvement over the conventional methods not accounting for initial defects that are inevitably present in all materials and structures. The potential of the fracture mechanics technology, however, has not been fully recognized. There remains much to be done in constructing a quantitative theory of material damage that can reliably translate small specimen data to the design of large size structural components. The work of the physical metallurgists and the fracture mechanicians should also be brought together by reconciling the details of the material microstructure with the assumed continua of the computational methods. It is with the aim of developing a wider appreciation of the fracture mechanics technology applied to the design of engineering structures such as aircrafts, ships, bridges, pavements, pressure vessels, off-shore structures, pipelines, etc. that this series is being developed.

Undoubtedly, the successful application of any technology must rely on the soundness of the underlying basic concepts and mathematical models and how they reconcile with each other. This goal has been accomplished to a large extent by the book series on *Mechanics of Fracture* started in 1972. The seven published volumes offer a wealth of information on the effects of defects or cracks in cylindrical bars, thin and thick plates, shells, composites and solids in three dimensions. Both static and dynamic loads are considered. Each volume contains an introductory chapter that illustrates how the strain energy criterion can be used to analyze the combined influence of defect size, component geometry and size, loading, material properties, etc. The criterion is particularly effective for treating mixed mode fracture where the crack propagates in a non-self similar fashion. One of the major difficulties that continuously perplex the practitioners in fracture mechanics is the selection of an appropriate fracture criterion without which no reliable prediction of failure could be made. This requires much discernment, judgement and experience. General conclusion based on the agreement of theory and experiment for a limited number of physical phenomena should be avoided.

Looking into the future the rapid advancement of modern technology will require more sophisticated concepts in design. The micro-chips used widely in electronics and advanced composites developed for aerospace applications are just some of the more well-known examples. The more efficient use of materials in previously unexperienced environments is no doubt needed. Fracture mechanics should be extended beyond the range of LEFM. To be better understood is the entire process of material damage that includes crack initiation, slow growth and eventual termination by fast crack propagation. Material behavior characterized from the uniaxial tensile tests must be related to more complicated stress states. These difficulties could be overcome by unifying metallurgical and fracture mechanics studies, particularly in assessing the results with consistency.

This series is therefore offered to emphasize the applications of fracture mechanics technology that could be employed to assure the safe behavior of engineering products and structures. Unexpected failures may or may not be critical in themselves but they can often be annoying, time-wasting and discrediting of the technical community.

Bethlehem, Pennsylvania   G.C. Sih
1984   Editor-in-Chief

Preface

In this volume on the mechanics of fracture of Portland cement concrete, the general theme is the connection between microstructural phenomena and macroscopic models. The issues addressed include techniques for observation over a wide range of scales, the influence of microcracking on common measures of strength and deformability, and ultimately, the relationship between microstructural changes in concrete under load and its resistance to cracking.

It is now commonly accepted that, in past attempts to force-fit the behavior of concrete into the rules of linear elastic fracture mechanics, proper attention has not been paid to scale effects. Clearly, the relationships among specimen size, crack length and opening, and characteristic material fabric dimensions have been, in comparison to their counterparts in metals, ceramics, and rocks, abused in concrete. Without a fundamental understanding of these relationships, additional testing in search of the elusive, single measure of fracture toughness has spawned additional confusion and frustration.

No one is in a better position to document this observation than Professor Mindess. Chuckles and white knuckles must have been his reward for his monumental effort in recently assembling an annotated bibliography with nearly 500 citations on the fracture of concrete, the vast majority in the area of physical testing. Professor Mindess effectively capsulizes this bibliography in his chapter, highlighting trends in classical linear and non-linear approaches to cracking in hardened paste, mortar, concrete, and its fiber reinforced and polymer impregnated versions. The reading here is sobering: How could so many have done so much for so long with so little practical results to show for the effort? One gets the feeling that if the average height of all the experimentalists had been about six meters, and their specimens scaled for no other particular reason accordingly, well, perhaps there would be a bit more agreement on how to measure toughness. Professor Mindess, like a good lawyer, summarizes, synthesizes, and redirects, and the direction is clear: We must meet the question of local damage head-on, not bury it under the rubble of additional test specimens.

If there are no plastic zones, shear lips, surface dimples or other local failure manifestations usually accompanying fracture in metals occurring in concrete, what damage mechanisms, what local energy sinks are occurring? The chapters by Professors Sih and DiTommaso both address this question by first examining the character of the response of concrete to common forms of loading. They conclude that microcracking, perhaps viewed over a number of different scales, is the fundamental source of non-linear response.

Professor DiTommaso treats the problem of damage accumulation in an inhomogeneous material by a gradation of approaches, first considering the response of a single microcrack in an infinite mortar matrix to mixed-mode loading. Step two is to include the certainty that some microcracking occurs at or near the interface of mortar to aggregate. Clearly, a matrix containing many cracks of arbitrary distribution and orientation is a step again more realistic. Finally, by use of the Monte-Carlo method, Professor DiTommaso puts it all together in a simulation of damage accumulation in a model containing mortar and a distribution of aggregates each with its own initial bond microcracks.

The driving force in the damage accumulation scheme of Professor Sih is the critical strain energy density. In an analysis that begs the reader for comparison with the approaches of Professors Bazant, Hillerborg, and Ingraffea (see volume 4 of this series), Professor Sih reproduces a complete load-displacement curve of a concrete beam in bending using this truly versatile approach. The reader should clearly trace the relationships among the maximum allowable strain in the critical strain energy density approach, the maximum COD in the fictitious crack model, and the blunt crack bandwidth in the Bazant model. Professor Sih's approach successfully captures the effect of specimen size on response, as well as producing distinct crack growth resistance curves for materials with different critical strain energy densities.

These latter two feats were also objectives for Professor Shah in his chapter. By using a modification to the fictitious crack model and without the necessity of finite element modeling, he predicts the effect of specimen geometry on the size of the crack tip damage zone. Double torsion, tapered double cantilever, and single-edge-notch bend geometries were analyzed in studying this effect as well as for generating crack growth resistance curves. The predictions for these are in good agreement with measured relationships.

The last two chapters of this volume differ in thrust from the others and in direction from each other. While both address physical testing techniques, that of Professor Slate peeks inside of concrete while the chapter by Dr. Jacquot is a primer on surface observations.

It is clear that techniques must be available for observing and quantifying the internal damage caused by microcracking that figures so importantly in the macro-models of the previous chapters. Professor Slate and Mr. Hover chronicle the development of X-radiography and microscope and dye methods for this purpose. Further, they relate the results of their observations to microcracking influences on a wide range of commonly measured properties, and to the influence of various loading states and rates on the damage process itself.

The reader is again reminded that all of the macro-models described in this volume set require some characteristic length parameter to sensitize them to scale effects. Maximum COD, r_0, or whatever, the state-of-the-art interferometry techniques being pioneered by Dr. Jacquot show promise for their direct measurement. Dr. Jacquot describes in detail the general classes of optical methods which might be applied to full-field surface measurements of the cracking process in concrete. Comparisons of the methods on the bases of sensitivities, ranges, error influences, and viewport area are made. Finally, Dr. Jacquot describes very recent, in some cases first time, experience with each of the methods for fracture mechanics studies, most on concrete itself.

The editors of this work never thought during the early stages of its development that it would grow to require two healthy volumes. It is a tribute to the quality of work of the contributors and a necessity born of the importance of the subject matter that there is so much to print. Savor it slowly and enjoy.

April 1983
Cornell University
University of Bologna

A.R. Ingraffea
A. Carpinteri

Contributing authors

A. DiTommaso
University of Bologna, Bologna, Italy

P. Jacquot
Swiss Federal Institute of Technology, Lausanne, Switzerland

S. Mindess
The University of British Columbia, Vancouver, British Columbia, Canada

S.P. Shah
Northwestern University, Evanston, Illinois

G.C. Sih
Lehigh University, Bethlehem, Pennsylvania

F.O. Slate and K.C. Hover
Cornell University, Ithaca, New York

Mechanics of material damage in concrete

1.1 Introduction

The loss of structure integrity in concrete due to cracking has attracted wide attention from engineers and researchers in recent years. One of the major advances in the field is the application of the fracture mechanics discipline. Concrete strength being sensitive to inherent flaws and aggregate composition will advertently be size dependent if these variables are accounted for in the analytical modeling. In addition, the rate of loading will also play a role as in the case of material behavior in general. The prerequisite for defining the thresholds of material damage necessitates the selection of suitable fracture or failure criteria.

Since controversy still prevails with regard to which fracture criterion can best describe the concrete behavior, current concepts involving displacement, stress and energy parameters are still not clear and often lead to inconsistencies. Keep in mind that material characterization as well as the translation of laboratory data to the design of full-size structures are essential. A sound knowledge of the underlying assumptions of each theory is therefore mandatory. In particular, the interaction of material inhomogeneity with loading rate should be identified with material damage at the appropriate scale level so that the analytical results can be assessed in terms of experimental data. The majority of available fracture mechanics theories apply the concept of fracture toughness for characterizing the fracture behavior of metal alloys, normally assumed to be isotropic and homogeneous. Whether these same assumptions could be carried over to the concrete structure that may be non-homogeneous will be the subject of discussion.

No attempt will be made to review the literature in the fracture and cracking of concrete for an excellent collection of bibliography on this subject has already been compiled and can be found in [1]. The past investigations appear to fall in two categories: one deals with the fracture toughness characterization of concrete by parameters born from the theory of linear elastic fracture mechanics (LEFM) and the other delves more on the morphology of cracks and fracture surfaces in an attempt to understand the influence of material inhomogeneity on cracking. The fracture resistance of concrete owing to different size, shape and volume fraction of aggregates has been the subject of many past investigations. There is a general trend to borrow the fracture mechanics technology developed for metal alloys and apply it directly to concrete. The same has happened in the field of fiber reinforced composite materials [2] where parameters such as the critical energy release rate G_c or critical stress intensity factors* K_c are frequently measured and widely reported. Their significance and usefulness, however, must be seriously questioned [3] since they depend sensitively on the relative orientation of load and fiber direction. The meaning of G_c or K_c in LEFM requires scrutiny for systems where the source of energy release may not coincide with the instability of a single crack. Additional energy can be dissipated during slow crack growth and local material damage before the catastrophic failure. Strictly speaking, the *fracture toughness* [4] K_{1c} in LEFM should be interpreted as a material behavior parameter† rather than a material constant. It simply refers to the event of sudden energy release during a unit extension of *homogeneous and isotropic material* at global instability. Material constants should be relatively insensitive to changes in specimen size, loading rate, etc. Standards and procedures for applying LEFM have been clearly established by ASTM [5] for metal alloys. Obvious conceptual, analytical and experimental modifications are necessary when other modes of material damage and/or slow crack growth occur. In concrete, breaking of the aggregates, cracking of the mortar, and/or debonding of mortar and aggregates can occur prior to global instability of the composite system. The order and damage extent of these different failure modes are load history dependent and should be addressed accordingly.

One way of modeling material damage by application of continuum mechanics discipline is to regard yielding as damage by microcracking and fracture by macrocracking. The combined effects of these two failure modes usually lead to non-linear response between load and displacement. Moreover, local failure and global instability no longer coincide and their distinction is essential in describing concrete failure. This involves the selection of an appropriate failure criterion. The theory of LEFM utilizing G_c or K_c and the concept of path independent integrals are basically inadequate because of their inability to recognize the intervening stage of subcritical material damage‡. The coincidence of local and global instability implies that all the stored energy is released instantly as in the case of ideal brittle fracture.

A consistent treatment of subcritical crack growth accompanied by local material damage in a three-point bending concrete specimen can be found in [6]. The concrete material is assumed to possess a bilinear softening relation between stress and strain. The initiation of slow crack growth was assumed to be associated with the local strain energy density function reaching a critical value $(dW/dV)_c$ and termination with the critical strain energy density factor S_c. The relation $r_c = S_c/(dW/dV)_c$ yields the size of the last ligament failure. To be consistent with the strain energy density criterion [7, 8], results reflecting the combined effects of loading increment, specimen size and fracture toughness are exhibited graphically by plotting the factor S with crack

---

*The notation K_c is used here to distinguish it from the ASTM valid K_{1c} definition [5]. It applies to a system that behaves linearly elastic up to fracture. Concrete does not satisfy this requirement.
†Other examples of material behavior parameters are the strain hardening coefficients for describing the nonlinear behavior of elastic-plastic materials.
‡Additional failure criteria need to be invoked to account for subcritical crack growth or damage. The application of several independent failure criteria for describing the same physical process lacks consistency and is prone to arbitrariness.

growth. The two extreme classical failure modes of plastic collapse and brittle fracture are identified as special cases of the more general treatment using $dW/dV = S/r$ as a unique failure criterion. The complete range of failure modes involving both yielding and fracture can thus be assessed quantitatively. It suffices to collect data from one set of loading increment and given size of specimen while the results for other loading increments and specimen sizes can be obtained from a S versus a plot.

In spite of the extensive research related to the influence of material inhomogeneity on structural component behavior, there is a widespread feeling that the subject is still not well understood. The slow progress in fundamental treatment of inhomogeneous materials is, no doubt, partly due to the lack of theoretical foundation concerning the interaction of microscopic entities with macroscopic parameters. The behavior of undamaged multi-phase media such as bodies containing spherical inclusions or voids [9, 10] can be treated in a straightforward manner. But one of the advantages of concrete is that its load carrying capacity is not greatly affected even when the aggregates and mortar may have damaged locally. The main concern is therefore to understand the behavior of partially damaged concrete systems and the mechanics of damage accumulation. The most serious obstacle to progress in this area may not necessarily be the complexity arising from stress analysis, but from determining the influence of subcritical material damage on global structural behavior. To this end, the concept of *local* and *global* stationary values of the strain energy density function dW/dV is introduced. A length parameter "*l*" can be calculated to measure the stability behavior of structural members by reflecting the combined effects of loading, geometry and material properties. Material and structure should be optimized and designed such that subcritical damage can be detected and remedied prior to catastrophic failure.

1.2 Mechanical strength of concrete

The properties of concrete are governed by a mixture of cement particles about 10^{-5} m in size, sand particles 10^{-4} m and coarse aggregates or small stones 10^{-2} m. The mortar consisting of cement gel and sands is usually regarded as a homogeneous and isotropic continuum while the whole system must be treated as a two-phase composite. Slow complex chemical reactions take place when water is added to the mixture as the system hardens into concrete. The resulting mechanical properties depend not only on the constituents but also on the conditions under which the concrete is set.

Two-phase system. The elastic modulus of concrete E_c can be formally estimated from the simple rule of mixture:

$$E_c = E_m V_m + E_a V_a \tag{1.1}$$

which assumes that the aggregates are bonded perfectly to the mortar or matrix. Such an idealized condition can never be met in practice as cracks are present at aggregate-mortar interfaces during the drying period. In equation (1.1), E_m and E_a are the elastic moduli, and V_m and V_a are the volume fractions. The subscript m and a are used to denote the mortar and aggregate. Measured values of E_a and E_m extend over a wide

3

range. Their limits are $3E_c < E_a < 10E_c$ and $0.5E_c < E_m < E_c$ while V_a can vary from 0.5 to 0.7. An empirical version of equation (1.1) can be given in the form

$$E_c = E_m(1 + \epsilon V_a) \tag{1.2}$$

The constant ϵ depends on the aggregate modulus E_a, size and shape. For V_a up to 0.6, ϵ is found to be approximately equal to 4.0. For a typical value of $E_m = 9,000$ MN/m² and $V_a = 0.6$, equation (1.2) gives $E_c = 30,600$ MN/m². The modulus of concrete is more than three times higher than that of the mortar.

Stress and strain behavior. Since concrete is weak in tension, most tests are performed in compression. Concrete has no linear elastic region in the stress-strain curve whose gradient decreases continuously until fracture, Figure 1.1(a). Loading and unloading do not follow the same path. Under controlled strain rate loading, the stress-strain curve decays quickly after reaching the peak stress, Figure 1.1(b). Although surface cracks have appeared on the specimen soon after the stress peaks, the specimen could still carry loads up to considerable strains before failure. The volume strain defined as $\Delta V/V$ also acquires a maximum at approximately the peak stress. With cracks present, however, the meaning of this parameter is not always clear.

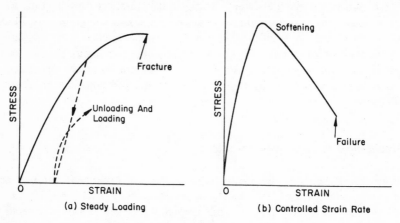

Figure 1.1.   Stress and strain curves for concrete.

Tensile tests have also been carried out. The three- or four-point bending of concrete beams are frequently used where the maximum fiber stress calculated at failure is taken as a measure of tensile strength. Another method is to split a right circular cylinder diametrically. A principal tensile stress acts in a horizontal plane along which the cylinder is assumed to fracture.

Strength consideration. The tensile strength of concrete is small but not negligible. It is of the order of $\sigma_c^{(t)} = 3.4$ MN/m² [11] and is less than that found for the mortar $\sigma_m^{(t)} = 6.87$ MN/m² [12, 13] and the coarse aggregate particles. In contrast to tension, the compressive strength of concrete $\sigma_c^{(c)} = 274$ MN/m² is higher than that for the mortar $\sigma_m^{(c)} = 103$ MN/m². The superscripts t and c are used to distinguish tension from compression. Much past efforts have been devoted to explain these discrepancies

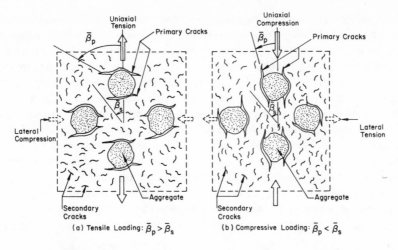

Figure 1.2.   Schematics of primary and secondary cracks in concrete.

in strength which are generally attributed to cracks that appear at the aggregate-mortar interfaces before loading is applied. This is caused by shrinking of the mortar matrix during drying out of the concrete. The bonding at interfaces is relatively weak such that cracking prevails at the aggregate interface. A consistent account of the following strength inequalities

$$\sigma_c^{(t)} < \sigma_m^{(t)}; \qquad \sigma_c^{(c)} > \sigma_m^{(f)}$$
$$\sigma_c^{(t)} < \sigma_c^{(c)}; \qquad \sigma_m^{(t)} < \sigma_m^{(c)} \tag{1.3}$$

is obviously needed to understand concrete behavior. The schematics displayed in Figures 1.2(a) and 1.2(b) are given to illustrate the different fracture patterns when loading is changed from tension to compression.

Strength variations in concrete are intimately associated with initial defects and crack growth. The mortar matrix contains inherent defects which are referred to as secondary cracks that are assumed to be oriented randomly making an average angle $\bar{\beta}_s$ with reference to the load axis, Figures 1.2(a) and 1.2(b). When a concrete specimen is stretched uniaxially, the interface cracks tend to close in the lateral direction while those aligned normal to tension tend to open, Figure 1.2(a). This effect becomes more and more pronounced as the load is increased until cracks are initiated from the interface. These are the primary cracks* and their orientations favor toward the planes normal to tension. On the average, they are aligned at an angle $\bar{\beta}_p$ with the load direction. On a probability basis, $\bar{\beta}_p$ is greater than $\bar{\beta}_s$. The situation is reversed during compression, Figure 1.2(b). The interface cracks in planes normal to compression tend to close and those parallel to compression tend to open because of lateral extension. The primary cracks emanating from the debonded portion of the interface now spread vertically parallel to the load axis. This is why $\bar{\beta}_p$ in compression becomes smaller than $\bar{\beta}_s$. The primary cracks dominate damage in concrete and the secondary cracks in mortar.

*The aggregates act simply as sites of crack initiation. Cracking of aggregates is neglected and would not affect the outcome of this part of the discussion qualitatively.

5

Having established the relative patterns of the primary and secondary cracks in concrete subjected to tension and compression, it is now possible to give a qualitative explanation of the strength inequalities given in equations (1.3). There exists a considerable amount of experimental evidence that the tensile and compressive strength of concrete-like materials are sensitive to the distribution and orientation of initial defects. More uniform distribution results in higher strength. The orientation with respect to load direction on the average can be measured through an angle, say $\bar{\beta}$. Based on the strain energy density criterion [14, 15], the tensile strength $\sigma^{(t)}$ was shown to decrease monotonically with this angle while the compressive strength $\sigma^{(c)}$ decayed rapidly for small $\bar{\beta}$ reaching a minimum and then increased as $\bar{\beta}$ is increased. This prediction [14] agreed well with fracture data on glass.

In order to establish a common basis of material damage reference, the results will be interpreted in terms of the single crack model. The solid curves in Figures 1.3(a) and 1.3(b) correspond to those for concrete and mortar specimens with a single crack

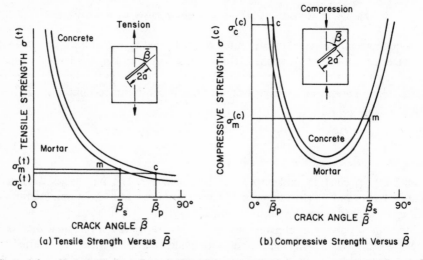

Figure 1.3.   Variations of tensile and compressive strength of concrete or mortar with angle $\bar{\beta}$.

of length, say 2a, put in artificially. The concrete is assumed to behave as a quasi-homogeneous system. The tensile strength $\sigma^{(t)}$ and compressive strength $\sigma^{(c)}$ are calculated as $\bar{\beta}$ is varied from $0°$ to $90°$. For the single crack model, both $\sigma^{(t)}$ and $\sigma^{(c)}$ for the concrete are higher than those for the mortar. This conclusion obviously does not hold if the cracking patterns in the specimens were to be developed naturally as indicated in Figures 1.2(a) and 1.2(b). In the case of uniaxial tension in Figure 1.2(a) for $\bar{\beta}_p > \bar{\beta}_s$, Figure 1.3(a) shows that $\sigma_m^{(t)}$ and $\bar{\beta}_s$ intersect at the point m which can be higher than the intersection of $\sigma_c^{(t)}$ and $\bar{\beta}_p$ at c. This explains why the tensile strength of concrete is slightly larger than that of mortar, i.e., $\sigma_m^{(t)} > \sigma_c^{(t)}$. The case of compression is shown in Figure 1.2(b) for $\bar{\beta}_p < \bar{\beta}_s$. The intersection of $\sigma_c^{(c)}$ and $\bar{\beta}_p$ at c is now much higher than that of $\sigma_m^{(c)}$ and $\bar{\beta}_s$ at m. This results in a higher compressive strength for concrete as compared with mortar, i.e., $\sigma_c^{(c)} > \sigma_m^{(c)}$.

The remaining two strength inequalities in equations (1.3) can be explained in the same way. Both points c and m in Figure 1.3(b) are seen to be higher than the corresponding points in Figure 1.3(a). The difference between the two m points, however, is smaller than that between the two c points. This confirms the experimental data [11−13] quoted earlier which correspond to $\sigma_m^{(c)}/\sigma_m^{(t)} = 14.99$ and $\sigma_c^{(c)}/\sigma_c^{(t)} = 80.59$. This trend is typical of materials weakened by the presence of initial defects that react differently in tension and compression. Experimental scatter between the ratio of compressive and tensile strength of concrete-like materials are expected if care is not given to account for initial defect orientations. There is no reason why a fixed ratio of $\sigma_c^{(c)}/\sigma_c^{(t)}$ should result as the value of $\bar{\beta}_p$ may vary from specimen to specimen. The value $\sigma_c^{(c)}/\sigma_c^{(t)} = 80.59$ quoted here is an order of magnitude different from the ratio $\sigma_c^{(c)}/\sigma_c^{(t)} = 8.0$ as claimed in [16]*. It is therefore not meaningful to compare results unless some reference is made to the initial state of defects in relation to loading direction.

1.3   Stress and fracture analysis

As mentioned earlier, concrete does not behave linearly elastic up to fracture and it is not really homogeneous and isotropic. Since many of the basic assumptions in LEFM are not satisfied, it is therefore necessary to examine how concrete fracture should be modeled. The nonlinear response of stress-strain shown in Figures 1.1 physically implies that concrete is being damaged by slow crack growth and/or local material yielding before catastrophic failure. Stress and failure analysis must therefore be performed for each increment of loading or material damage. Regardless of whether the analytical model is one of multiple cracks or a single dominant crack, the stage of subcritical material damage cannot be ignored†.

The choice of modeling the concrete as a nonhomogeneous medium containing discrete numbers of aggregate particles or a quasi-homogeneous material with some combined average properties of the aggregate and mortar matrix depends on the difference between E_a and E_m and the volume fraction V_a of the aggregates. A discussion on this choice for the fiber reinforced composite materials is given in [17]. The basic idea applies equally well to concrete or any two-phase systems. Stress analysis must be performed in conjunction with a suitable failure criterion that can consistently account for the physical material damage process.

Two-phase nonhomogeneous system. If the aggregate volume fraction is relatively low, then the individual failure modes of aggregate breaking, mortar cracking and interface debonding should be identified in terms of the global load and displacement measurements. It suffices to know the fracture toughness of the aggregate and mortar for estimating the amount of energy dissipated through cracking. Debonding at aggregate-mortar interface is more complex to model as it can exist in loaded and unloaded concrete, Figure 1.4. Microcracks are frequently observed in a local region of high material porosity. Such a region may be modeled as macroyielding that leads to

---

*This result was not based on the Griffith criterion. The ratio $\sigma_c^{(c)}/\sigma_c^{(t)} = 8.0$ is a consequence of the local maximum normal stress criterion.
†The LEFM fracture toughness characterization technique based on G_c or K_c leaves out subcritical crack growth.

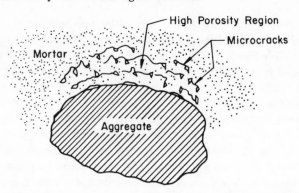

Figure 1.4.   Debonding at aggregate-mortar interface neighborhood.

eventual macrocracking or debonding*. Macrocrack initiation and growth must be addressed.

A detailed treatment of concrete as a nonhomogeneous system leaves no room for the single parameter G_c or K_c type of consideration. The objective is then to identify the state of local material damage with the loading capacity of the concrete in terms of the global load and displacement parameters.

Quasi-homogeneous system. For sufficiently high volume fraction of the aggregates and controlled loading rates such that the primary damage of concrete can be characterized by a slowly growing crack while microcracking may be modeled as macroyielding, the quasi-homogeneous assumption can lead to reasonable predictions. The controlled strain rate loading of a three-point bending specimen with an initially inserted crack is such an example. Even though material inhomogeneity have been relaxed, the evaluation of specimen size and loading rate effects being typical of concrete behavior is still non-trivial. The appropriate selection of failure criterion and execution of stress analysis remain as the prerequisites for analyzing material damage behavior.

Material damage criterion. The majority of criteria either address global and local failure separately or assume that they coincide. Most of the physical material damage process, however, starts with initiation, subcritical growth and final termination. These three stages are all observable at the macroscopic scale level and hence they should be all included in a continuum mechanics model. Assuming that failure modes could be uniquely associated with threshold values of energy state in a unit volume of material, damage initiation and growth can then be consistently accounted for by a single failure criterion. This quantity is the strain energy density function[†]

$$\frac{dW}{dV} = \int_0^{\epsilon_{ij}} \sigma_{ij} d\epsilon_{ij} \tag{1.4}$$

---

*Cohesive failure is usually assumed because of the inability to define or model a real interface [18].

[†]The form of equation (1.4) applies to isothermal problems. Otherwise, an additional function depending on temperature change must be added.

that applies to all materials and geometric discontinuities within the framwork of continuum mechanics. The stress and strain components are denoted by σ_{ij} and ϵ_{ij}. The critical value of dW/dV or $(dW/dV)_c$ is the area under the true uniaxial stress and strain curve at fracture. It has been measured and reported for a number of metal alloys [19, 20]. Failure initiation is thus assumed to coincide with dW/dV in a material element reaching $(dW/dV)_c$ while damage accumulation or growth is assumed to follow the relation

$$\left(\frac{dW}{dV}\right)_c \quad \text{or} \quad \left(\frac{dW}{dV}\right)_c^* = \frac{S_1}{r_1} = \frac{S_2}{r_2} = \ldots = \frac{S_j}{r_j} = \text{const.} \tag{1.5}$$

in which S_j are the strain energy density factors and r_j the growth increments $j = 1, 2, \ldots, n$. This process may lead to damage arrest or global instability of the system depending on whether S_j/r_j approaches S_0/r_0 or S_c/r_c. Damage may involve plastic collapse, brittle fracture or a combination of yielding and fracture. If the final form of damage corresponds to sudden fracture, then the ligament of material at incipient fracture is given by

$$r_c = \frac{S_c}{(dW/dV)_c} \tag{1.6}$$

The critical value S_c can be related to the commonly known fracture toughness value K_{1c} [14]:

$$S_c = \frac{(1 + \nu)(1 - 2\nu)}{2\pi E} K_{1c}^2 \tag{1.7}$$

where ν is the Poisson's ratio and E the Young's modulus. Values of S_c for many engineering materials are given in [21].

Frequently, material elements are damaged by yielding before they break. In other words, some of the stored energy, say $(dW/dV)_p$, is dissipated prior to material separation. Therefore, only the amount

$$\left(\frac{dW}{dV}\right)_c^* = \left(\frac{dW}{dV}\right)_c - \left(\frac{dW}{dV}\right)_p \tag{1.8}$$

is available for release at the instant of fracture. In that case, $(dW/dV)_c^*$ in equation (1.5) must be used instead of $(dW/dV)_c$.

1.4 Damage analysis of concrete beam in bending: effects of softening and loading step

Concrete is known to behave differently when specimen size and/or loading increment are changed. These effects interact with variations in material properties and are difficult to sort out as they further depend on the presence of defects that may or may not grow during loading. Material damage in concrete is recognized as a load-history dependent process that must be treated accordingly.

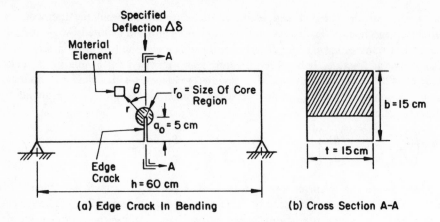

(a) Edge Crack In Bending   (b) Cross Section A-A

Figure 1.5.   Three-point bending of concrete beam.

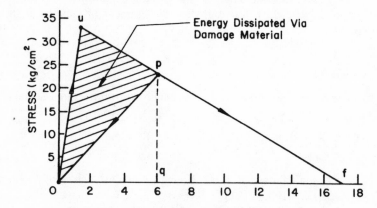

Figure 1.6.   Bilinear stress strain curve for concrete with softening.

Concrete can be made to change its behavior from brittle to ductile simply by altering the loading increment in addition to changes in its composition. Such changes are reflected through the critical stress intensity factor K_c [22] in LEFM. Numerous past investigations [23–27] have been devoted to study the influence of crack growth on concrete behavior. However, a complete description of the initiation of slow crack growth leading to global instability is still lacking even for the single crack model. Unless this material damage phenomenon is assessed quantitatively, no confidence can be placed in translating test data to design.

Material damage model. The three-point bending specimen with an edge crack will be depicted for this discussion [6]. A typical set of beam and crack dimensions are shown in Figure 1.5. Each material element is assumed to behave in a quasi-homogeneous* fashion following the stress and strain relation in Figure 1.1(b) when the beam is loaded incrementally by specifying constant deflection steps $\Delta\delta$. As a good approximation, the bilinear stress-strain relation in Figure 1.6 will be adopted in the analysis

---

*Henceforth, the subscript c on the modulus E for concrete will be dropped since no distinctions among the moduli for aggregate, mortar and concrete are necessary in the quasi-homogeneous system.

to follow. The stress will increase linearly with strain up to the point of ultimate strength u whose coordinates are ϵ_u and σ_u. Thereafter, the strain continues to increase to ϵ_f at f while the stress decreases linearly to zero, i.e., $\sigma_f = 0$. At any point p, the material experiences a reduction in the modulus from E to E^* corresponding, respectively, to the slopes of the lines ou and op. Irreversibility due to material damage is invoked by unloading along the line po. The degradation of moduli for elements near the crack tip in Figure 1.5(a) will occur nonuniformly depending on the position of the point p between u and f. The relation between σ and ϵ at p can be expressed as

$$\sigma = E^* \epsilon = \frac{\sigma_u(\epsilon_u + \epsilon_f)}{\epsilon_f + (\sigma_u/E^*)} \tag{1.9}$$

where ϵ_f is measured from the point u. The effective modulus E^* is discretized to 25 different steps:

$$E^*(n) = \frac{26 - n}{25} E, \qquad n = 1, 2, \ldots, 25 \tag{1.10}$$

which is adjusted in accordance with the current value of dW/dV obtained from the stress analysis for each increment of crack growth:

$$\frac{dW}{dV} = \frac{1}{2}(\sigma\epsilon + \sigma_u\epsilon - \sigma\epsilon_u) \tag{1.11}$$

This scheme of material damage-accumulation was used in [28] to analyze metal failure by yielding and fracture. The only difference is that the metal followed a strain hardening behavior instead of softening as in the case of concrete.

The permanent damage in a material element is measured by the area oup which corresponds to $(dW/dV)_p$ in equation (1.8), i.e.,

$$\left(\frac{dW}{dV}\right)_p = \frac{1}{2}(\sigma_u\epsilon - \sigma\epsilon_u) \tag{1.12}$$

The recoverable energy density corresponds to the area opq. It follows that the available energy density for sudden release at fracture is the sum of the areas opq and pfq, i.e.,

$$\left(\frac{dW}{dV}\right)^* = \frac{1}{2}(\sigma_u\epsilon_u + \sigma_u\epsilon_f - \sigma_u\epsilon + \sigma\epsilon_u) \tag{1.13}$$

In other words, the elements local to the crack tip region can be damaged before they break. The available energy for release during a segment of crack growth is not the total amount (dW/dV) but $(dW/dV)^*$.

Material properties and loading steps. Referring to Figure 1.6, all concretes behave similarly up to the point u with $E = 36.5 \times 10^4$ kg/cm^2, $\nu = 0.1$, $\sigma_u = 31.9$ kg/cm^2 and $\epsilon_u = 0.87 \times 10^{-4}$. The softening portion of the curve from u to f can vary. Table I refers to three different types of softening behavior distinguished as Materials A, B and C. The values of E^* and $(dW/dV)_c^*$ corresponding to the amount of absorbed strain

11

Table I.   Three different softening rates for concrete.

Material Type	Softening Strain ϵ_f (cm/cm $\times$ 10^{-4})	Critical Strain Energy Density $(dW/dV)_c$(kg/cm^2 $\times$ 10^{-3})
A	16.0	26.90
B	8.0	14.14
C	4.0	7.7

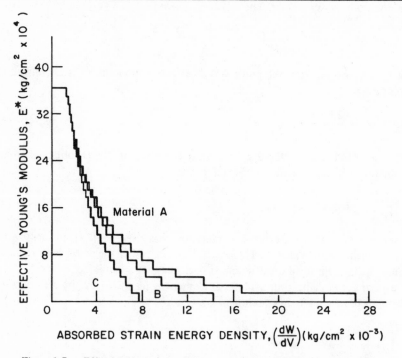

Figure 1.7.   Effective Young's modulus versus absorbed strain energy density.

energy density are given in Figures 1.7 and 1.8 for the three different materials. The severity of material damage increases with increasing dW/dV absorbed in a given element. This translates into a degradation in E^*, Figure 1.7, and less available energy for release at fracture, Figure 1.8.

The beam will be loaded by a constant deflection increment $\Delta\delta = 4 \times 10^{-3}$ cm while the material properties are varied according to Table I. Only the type C material will be subjected to two other deflection increments of $\Delta\delta = 2 \times 10^{-3}$ and 1×10^{-3} cm. The size of the beam will be kept constant with $b = t = 15$ cm and $h = 60$ cm in this part of the investigation. Specimen size effect will be analyzed subsequently by varying the dimension b.

Crack growth analysis. Starting with an initial crack of length $a_0 = 5$ cm as indicated in Figure 1.5(a), the beam is bent by increasing the deflection in constant steps of $\Delta\delta$. Material elements near the crack tip will first damage by yielding and then break

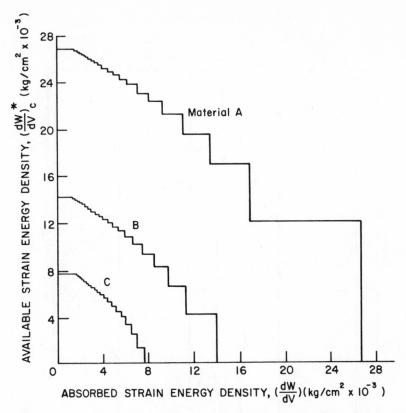

Figure 1.8.   Available strain energy density versus absorbed strain energy density.

to initiate crack growth following the relation in equation (1.5). Stress analysis is performed for each increment of crack growth by application of the Axisymmetric/ Planar Elastic Structures (APES) finite element computer program [29]. The grid pattern in Figure 1.9 is generated by using 52 elements and 309 nodes. Use is made of the twelve nodes quadrilateral isoparametric elements that result in cubic displacement and quadratic stress and strain fields. Embedded into the crack tip region is the $1/r$ singular strain energy density field by means of the 1/9 to 4/9 nodal spacing on elements adjacent to the crack edge.

For each combination of material and loading step, different crack growth steps a_j ($j = 1, 2$, etc.) are taken until the crack has grown more than sixty percent through the cross section. The degradation of the local moduli E^* in the neighboring crack tip elements is measured by n from 1 to 25. Material damage increases with n. The undamaged elements are not numbered while those damaged are referenced by the Roman numerals I, II, ... , X as given in Figure 1.9. A glance at the tabulated values of n in Table II shows that the elements I, II, III, etc., close to the crack tip are damaged more severely as the loading step is increased. Material A being the toughest sustained more incremental crack growth while material C being the weakest had the least number of crack growth steps.

Illustrated in Figure 1.10 is a detail account of incremental loading for Material A

13

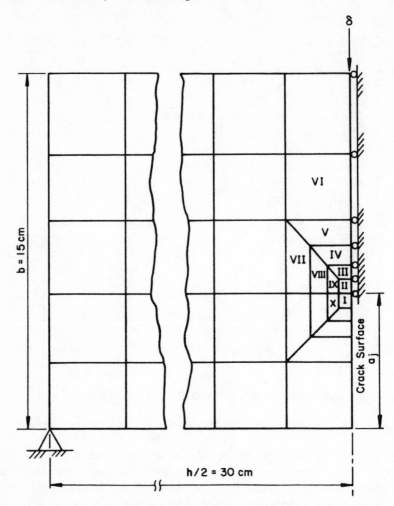

Figure 1.9.   Finite element grid pattern.

with $\Delta\delta = 4 \times 10^{-3}$ cm. The relationships between P and δ for the six crack growth steps corresponding to those in Table II are shown. For each increment of deflection $\Delta\delta$, the specimen is loaded to point C obtained by connecting the origin to the point on the solid curve established from the previous load increment. The loss in stiffness due to local material damage and crack growth then determines the position of the current point E. This process is repeated while the specimen accumulates damage and becomes more and more flexible. The ratio

$$\frac{CD}{CE} = \frac{\text{Stiffness loss by material damage}}{\text{Total stiffness loss}} \tag{1.14}$$

represents the relative stiffness loss due to local material damage ahead of the crack. Hence, $1 - (CD/CE)$ is the corresponding loss in stiffness due to crack growth. During the first load increment, material damage is confined to the crack tip region and the

Table II.   Crack growth and local material damage data for Materials A, B and C with $\Delta\delta = 4 \times 10^{-3}$ cm.

Deflection δ_j(cm × 10⁻³)	Load P_j (kg)	Crack Length a_j (cm)	Damage Element Reference Number									
			I	II	III	IV	V	VI	VII	VIII	IX	X
Material A (j = 1, 2, . . . , 6)												
4	574	5.088	7	18	4	–	–	–	–	·	4	–
8	967	5.177	21	24	22	11	–	–	–	10	18	9
12	996	5.653	25	25	25	23	15	–	9	19	23	12
16	706	6.999	25	25	25	25	23	9	13	21	25	21
20	400	8.666	25	25	25	25	24	14	13	21	25	18
24	238	10.269	25	25 -	25	25	24	1	13	21	25	21
Material B (j = 1, 2, . . . , 4)												
4	564	5.157	7	19	4	–	–	–	–	–	5	–
8	828	5.637	23	25	24	12	–	–	–	11	20	9
12	572	7.325	25	25	25	24	17	–	9	20	23	9
16	276	9.325	25	25	25	25	22	6	9	21	25	9
Material C (j = 1, 2, 3)												
4	528	5.420	8	22	4	–	–	–	–	–	5	–
8	582	6.828	25	25	25	14	–	–	–	12	23	7
12	287	9.000	25	25	25	25	17	–	6	21	25	7

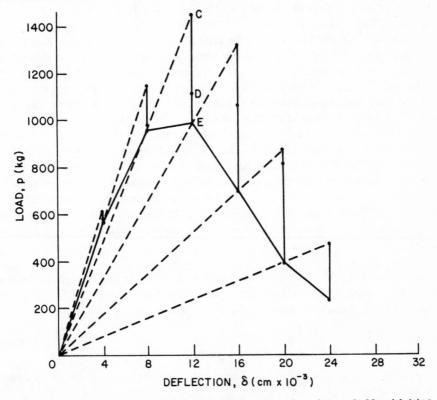

Figure 1.10.   Load-displacement curve for material damage and crack growth: Material A ($\Delta\delta = 4 \times 10^{-3}$ cm).

Table III.   Damage accumulation data for Material C with $\Delta\delta = 2 \times 10^{-3}$ and 1×10^{-3} cm.

Deflection δ_j (cm $\times 10^{-3}$)	Load P_j (kg)	Crack Length a_j (cm)	Damage Element Reference Number									
			I	II	III	IV	V	VI	VII	VIII	IX	X
$\Delta\delta = 2 \times 10^{-3}$ cm ($j = 1, 2, \ldots, 5$)												
2	302	5.069	−	2	−	−	−	−	−	−	−	−
4	521	5.490	8	22	4	−	−	−	−	−	5	−
6	566	6.299	21	25	23	6	−	−	−	3	16	−
8	384	7.832	25	25	25	22	2	−	−	9	17	−
10	166	9.832	25	25	25	25	12	−	−	9	17	−
$\Delta\delta = 1 \times 10^{-3}$ cm ($j = 1, 2, \ldots, 8$)												
1	152	5.023	−	−	−	−	−	−	−	−	−	−
2	301	5.092	−	2	−	−	−	−	−	−	−	−
3	426	5.258	−	15	−	−	−	−	−	−	−	−
4	483	5.702	8	25	8	−	−	−	−	−	2	−
5	500	6.208	18	25	22	2	−	−	−	−	8	−
6	392	7.318	25	25	25	13	−	−	−	−	8	−
7	273	8.526	25	25	25	20	−	−	−	−	8	−
8	119	10.155	25	25	25	25	−	−	−	−	8	−

specimen stiffness is reduced only by approximately six percent. Nonlinearity begins at the second step while a maximum load of $P_3 = 996$ kg is reached at the third step with $\delta_3 = 12 \times 10^{-3}$ cm. Thereafter, the tangent modulus becomes negative with decreasing load. At the fourth step, the ratio CD/CE is nearly one-half. This means that the decrease in secant modulus due to material damage and crack growth is about the same. As is to be expected, crack growth begins to dominate as the beam deflection increases. The ratio CD/CE or DE/CE changes for each load step and is sensitive to the manner with which the beam is loaded incrementally. This exhibits the load-history dependent character of material damage and crack growth.

In the same way, Material C is analyzed for two additional deflection steps of $\Delta\delta = 2 \times 10^{-3}$ and 1×10^{-3} cm. Owing to the fact that Material C has a relatively low toughness and the deflection steps have been reduced, the crack tip region is not damaged as severely as those cases given in Table II. Referring to the element number system I, II, $\ldots$, X in Figure 1.9, the results are summarized in Table III. The elements VI, VII and X are undamaged. Load and displacement relationships similar to that given in Figure 1.10 can be developed to display the different degree of material damage.

Damage associated with yielding. It is of interest to compare the locations of damage zone with yielding predicted from the strain energy density criterion [14, 15]. Let damage in the jth element be measured by the reduction in the elastic modulus as

$$d_j = \frac{E - E_j^*}{E} \tag{1.15}$$

With reference to the centroids (x_j, y_j) of the element, the center of the damage zone is located by

$$x_d = \sum_j \frac{x_j d_j A_j}{\sum_j d_j A_j}, \qquad y_d = \sum_j \frac{y_j d_j A_j}{\sum_j d_j A_j} \qquad (1.16)$$

where A_j is the area of an element. These locations for Material C subjected to the three different deflection steps are plotted in Figure 1.11 with reference to the crack tip position. They tend to congregate off to the side of the crack.

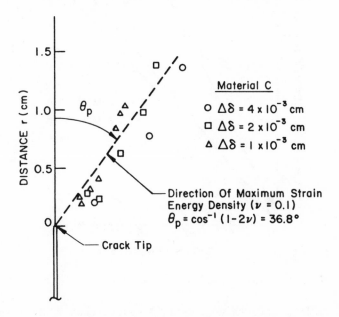

Figure 1.11   Local damage associated with yielding.

By using the asymptotic elastic stress solution, the strain energy density factor S can be calculated as a function of θ measured from the line of expected crack growth which corresponds to the vertical line or crack plane in Figure 1.11. The relative minimum S can be shown to correspond with $\theta_0 = 0°$ and relative maximum with $\theta_p = \cos^{-1}(1-2\nu)$. According to the S-criterion [14,15], the former refers to the direction of macrocrack growth and the latter to the direction of maximum yielding. For $\nu = 0.1$, an angle of $\theta_p = 36.8°$ is found. The agreement is quite good for small values of r. This is to be expected as the asymptotic stress solution is limited to $r/a <$ 1/10. Yielding can thus be viewed as damage with loss of material stiffness. A similar conclusion was also obtained for metal in [28].

Resistance curves. Damage resistance curves due to changes in material toughness and loading step are obtained by plotting the strain energy density factor S as a function of the crack length a. Figure 1.12 gives the results for Materials A, B and C as specified in Table I. The condition $dS/da = $ const. is seen to prevail. The curve for Material A with the highest $(dW/dV)_c$ or fracture toughness has the larger slope. This implies that for the same energy input, a small crack would be inflicted in the tougher material. These results are consistent with the experimental observation on materials for varying

17

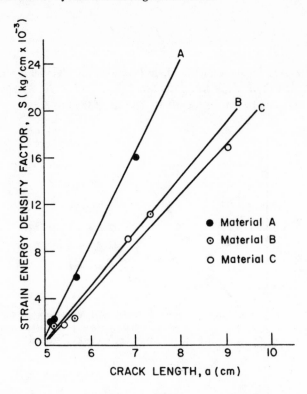

Figure 1.12   Resistance curves for Materials A, B and C with $\Delta\delta = 4 \times 10^{-3}$ cm.

degrees of ϵ_f. In general, the slope of the S versus a curve is directly proportional to ϵ_f. A slow crack growth threshold strain energy density factor S_0 also appears to occur at the point where all the curves in Figure 1.12 intersect. The quantity dS/da possesses the same units as $(dW/dV)_c$ and the ratio $dS/da/(dW/dV)_c$ is always less than unity. This ratio can serve as a relative measure of material damage to crack growth. In the limit as $dS/da/(dW/dV)_c$ tends to unity, the structure will fail strictly by brittle fracture with no prior damage of material as in the classical linear elastic fracture mechanics theory.

The effect of loading steps can also be exhibited by plotting the strain energy density factor S as a function of crack length a. This is given in Figure 1.13 for Material C with $\Delta\delta = 4 \times 10^{-3}$, 2×10^{-3} and 1×10^{-3} cm. The slope dS/da increases as $\Delta\delta$ is increased. This means that for a given material with $S_c = \text{const.}$, the critical crack length decreases if large loading steps are taken. This result is important for assessing the loading capacity of concrete in terms of the interaction of loading steps with critical flaw size. The trend of the curves in Figure 1.13 corresponds to that observed experimentally.

1.5 Scaling of specimen size

It is well-known that the smaller specimens tend to fail by plastic collapse while the large specimens can store sufficient energy to be released suddenly and fail by rapid

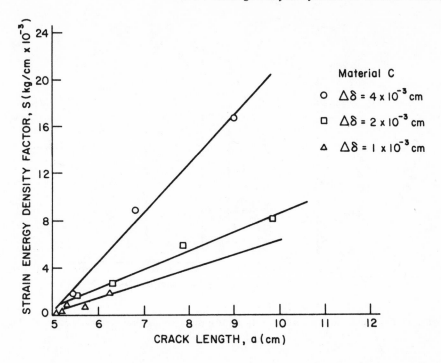

Figure 1.13.   Resistance curves for Material *C* with three different deflection steps.

fracture. Therefore, specimen size alone can alter the modes of failure [30]. Failure modes corresponding to varying degrees of yielding and crack growth are difficult to analyze because their behavior depends on load history. Moreover, crack growth takes place in a non self-similar fashion. The height of the beam *b* in Figure 1.5 will be varied by factors of 1, 2 and 3 such that the ratio $\Delta\delta/b = 2.6 \times 10^{-4}$ is kept constant. Summarized in Figure 1.14 are the relations between $P/(dW/dV)_c b^2$ and δ/b for $b = 15, 30, ---, 222$ cm. The vertical lines with the arrows indicate the limiting values of δ/b as the critical strain energy density value of $S_c = 8 \times 10^{-3}$ kg/cm is reached. This corresponds to $K_c = 143.36$ kg/cm$^{3/2}$ which is typical for concrete [31]. It is clear that crack instability occurs for smaller deflection of the specimen as the size *b* is increased. Without considering unstable crack propagation, the maximum load can be estimated from the relation

$$P_{\max} = 332.90 \left(\frac{dW}{dV}\right)_c b^2 \qquad\qquad (1.17)$$

Shown also in Figure 1.14 is that structural instability or collapse occurs before unstable crack propagation only in the case of $b = 15$ cm. For $b = 30$ cm, softening behavior is not present and the crack started to spread in an unstable manner when the load *P* is still in the ascending stage. The critical crack length a_c for each case can be found from the *S* versus *a* plots in Figure 1.15. Complete brittle fracture behavior is obtained when $b = 222$ cm. Specimen size effect corresponds to translation of $dS/da = $ const. lines.

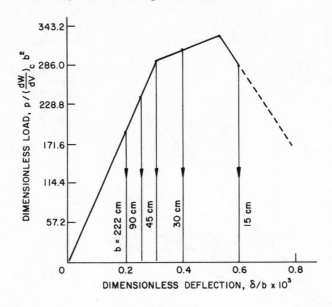

Figure 1.14.   Dimensionless load-deflection for unstable crack propagation in Material C with $\Delta\delta/b = 2.6 \times 10^{-4}$.

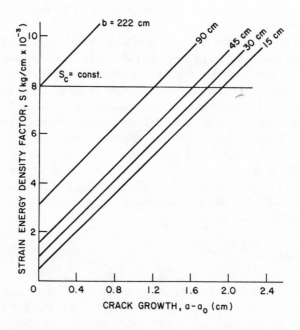

Figure 1.15.   Strain energy density factor versus crack length for Material C with different size scales and $\Delta\delta/b = 2.6 \times 10^{-4}$.

Of interest is to ~~prepare~~ compare the results with that suggested by ASTM [5]. The maximum load $P^{(2)}_{max}$ at fracture for different values of b can be found from the formula

$$K_1 = \frac{P^{(2)}h}{tb^{3/2}}\, f(a_0/b) \tag{1.18}$$

in which the function $f(a_0/b)$, stands for

$$f\left(\frac{a_0}{b}\right) = 2.9\left(\frac{a_0}{b}\right)^{1/2} - 4.6\left(\frac{a_0}{b}\right)^{3/2} + 21.8\left(\frac{a_0}{b}\right)^{5/2} - 37.6\left(\frac{a_0}{b}\right)^{7/2} + 38.7\left(\frac{a_0}{b}\right)^{9/2} \tag{1.19}$$

The predicted values of $P^{(3)}_{max}$ from a perfectly-elastic limit analysis are also computed according to

$$P^{(3)}_{max} = \frac{2}{3}\frac{\sigma_u t(b-a_0)^2}{h} \tag{1.20}$$

where the ligament size at collapse is assumed to be $b - a_0$. Normalizing the results obtained by the strain energy theory $P^{(1)}_{max}$ and $P^{(3)}_{max}$ from equation (1.20) with that of $P^{(2)}_{max}$ recommended by ASTM and plotting the results against b, Figure 1.16 summarizes the results. The line $P^{(1)}_{max}/P^{(2)}_{max}$ equal to 100% represents the limiting case of

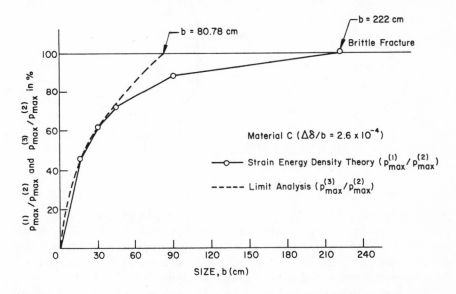

Figure 1.16.   Comparison of normalized maximum load ratio for brittle fracture and limit analysis with the strain energy density theory.

ASTM where failure coincides totally with brittle fracture. The curve in Figure 1.16 gives the maximum failure load involving both structure collapse and brittle fracture. Limit analysis predicts a transition b value of 80.78 cm while the present model yields $b = 222$ cm. The different failure modes between plastic collapse and brittle fracture have thus been assessed quantitatively by application of the strain energy density theory.

1.6 Long time behavior and failure of concrete

Failure by yielding and fracture in concrete is considered to be a short-time behavior. The stress and strain response can differ dramatically when the loading rate is altered. If the load level is decreased sufficiently and applied over a long period of time, concrete can behave very differently. This is referred to as creep and the test is usually carried out in compression. Creep of concrete is a complex process that is not yet fully understood. Under normal compression tests, the strain tends to rise quickly and then approaches a final value after some period of time. Several mechanisms are involved with the interaction of load distribution and material composition.

Creep behavior. When a compressive load is applied to a concrete specimen, the variations of strain with time consist of several segments as shown schematically in Figure 1.17. The strain that follows the sudden application of load is practically

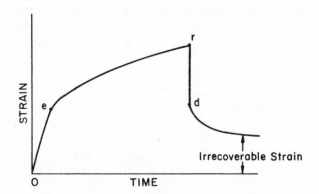

Figure 1.17.   Schematic of concrete behavior in creep.

instantaneous up to the elastic limit *e*. This is followed by an increase in strain over periods of months or years. The curve tends to a limiting value. If the load is suddenly removed at *r*, there is a large abrupt decrease of strain to the point *d*. The drop from *r* to *d* represents the instantaneous elastic recovery. A gradual tapering corresponds to the delayed elastic recovery leading to an asymptotic irrecoverable strain.

The observed creep behavior in Figure 1.17 has been attributed to the flow of adsorbed water out of the cement gel pore and capillaries as a result of external load. Other factors such as the closure of internal voids, plastic or viscous flow in the hardened cement paste and deformation of the aggregates have also been considered.

There is much controversy in explaining whether lowering of the aggregate modulus would tend to increase the creep deformation of concrete.

Loading and material inhomogeneity. Creep phenomena are much more difficult to analyze and quantify. As the deformation rates are slowed down, there is ample time for the induced stress to react with material inhomogeneity at a lower scale level. Greater attention must therefore be focused on redistribution of stress or energy state in disordered regions such as pores or defects in the mortar matrix and cracks at the aggregate-mortar interface. Creep rate is known to depend nonlinearly on porosity in the cement which may vary from location to location. A body having a mixture of more porous and less porous regions can have a different creep rate than one that has the same amount of porosity but distributed more homogeneously. The surface roughness of the aggregates can also affect the creep rate of concrete. Irregular surfaces tend to enhance interlocking and provide more resistance to creep. Another important factor is the pressure difference among the cement gel pores created by loading that promotes slow movement of water. In long time creep tests, the rate of water evaporation can significantly alter the creep behavior.

Creep within the concrete being nonhomogeneous acquires preferential direction depending on the energy state of spatial distribution of porosity and/or defects. Rearrangement of the preferential direction can occur as a function of time and nonuniformly due to change in damage pattern caused by the previously experienced energy state. For instance, microcracks can develop in regions of high porosity near the aggregate-mortar interface, Figure 1.4. The interaction between stress or energy state with material inhomogeneity arising from difference in material properties and damage patterns cannot be ignored. Since most concretes are tested in compression or in bending, different results are likely to prevail for creep in tension or combined loadings. It is apparent that porosities or voids tend to open more readily in tension than in compression. As a consequence, creep rate is expected to increase faster in a tensile test than in compressive test.

Analytical modeling. Presently, there is no adequate theoretical treatment of creep deformation for concrete. The viscous creep behavior assumed in many of the mathematical models neglects any change of the stress or energy state on a unit volume of material whose transient effect may not be negligible. If the creep process in a body leads to gradual change of the energy state in the volume elements, the transient component may not be negligible even for relative slow rates of deformation. The rearrangement of porosity and/or defect distribution also depends on time and contributes to the change in energy density. Since any irreversible material damage process such as creep is stress-strain history dependent, the current energy state is an integral part of material damage in the next increment of loading. Stress and damage analysis must be performed in tandem for each increment of the creep deformation process. Oversimplification can lead to models that are so remote from realities as to make them of little practical value.

1.7 Local and global stationary values of strain energy density function

In developing analytical models to describe material damage, it is imperative to make them sufficiently realistic and reduce them to manageable proportions. One of the major difficulties in assessing load interaction with nonhomogeneity due to material and physical damage is to express their combined effects quantitatively. This can be accomplished analytically from the local and global stationary values of the strain energy density function dW/dV. This new concept has been applied to a simple non-homogeneous material system involving a single inclusion [32] and will only be introduced briefly.

Stationary values. When a continuum is stressed, the energy stored in the volume elements vary from point to point and can undergo fluctuation in time as well. This variation depends on the combined effects of loading rate, continuum size and shape and material inhomogeneity*. A unique feature of all problems with or without initial defects is that there exists a local maximum and minimum of dW/dV at each point o_j ($j = 1,2,---,n$) confined to the *interior* of the domain. The maximum values of these local stationary values will be denoted by $[(dW/dV)_{\max}^{\max}]_l$ and $[(dW/dV)_{\min}^{\max}]_l$. Their stationary values are referred to the space variables r_j and θ_j ($j = 1,2,---,n$) in Figure 1.18(a) for a two-dimensional system. For simplicity, r_j are kept constant as $(dW/dV)_j$ are considered to vary with θ_j, Figure 1.18(b). In accordance with the

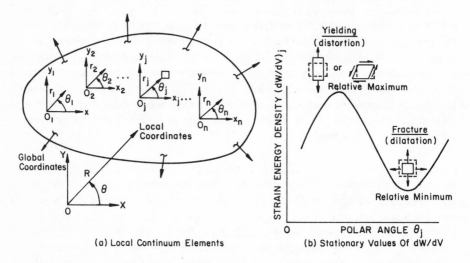

(a) Local Continuum Elements (b) Stationary Values Of dW/dV

Figure 1.18.   Fluctuations of strain energy density function in a continuum.

strain energy density criterion [14,15], the peaks (or maxima) can be related to the sites of yield initiation and the valleys (minima) with the sites of fracture initiation. This interpretation holds for material elements in the elastic as well as the plastic region** [33]. In addition, there also exists a corresponding pair of global stationary values of dW/dV with reference to the coordinates R and θ, Figure 1.18(a). They will generally occur at different locations and will be designated as $[(dW/dV)_{\max}^{\max}]_g$ and $[(dW/dV)_{\min}^{\max}]_g$. The fact that the locations of the local and global stationary values of dW/dV differ implies that local and global failure cannot occur simultaneously. The difference between $[(dW/dV)_{\min}^{\max}]_l$ and $[(dW/dV)_{\min}^{\max}]_g$ is measurable through

---

*Material inhomogeneity guarantees fluctuation in dW/dV however small for any loading systems.
**Additional distortion in already yielded material may be regarded as second degree of yielding [8].

a length parameter *l* which is sensitive to changes in loading rate, specimen geometry and material inhomogeneity. In the sequel, only the local and global minima of dW/dV will be discussed in connection with fracture initiation. The same applies to yield initiation when the maxima of local and global of dW/dV are used.

Local and global instability. In order to be specific, simple examples will be cited to explain the meaning of *l*. Consider the two systems in Figures 1.19(a) and 1.19(b)

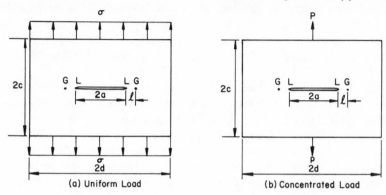

Figure 1.19.   Interpretation of local global instability.

that correspond to a center crack panel subjected, respectively, to a uniform stress $\sigma = 7\,\text{MN/m}^2$ and concentrated load $P = 4.45\,\text{kN}$. For $c = 6.35\,\text{cm}$ and $d = 12.7\,\text{cm}$, the length *l* is computed for $a = 2.54, 5.08, 7.62$ and $10.16\,\text{cm}$ while $\nu = 0.3$ and $E = 206.84 \times 10^3\,\text{MPa}$. The results are summarized in Table IV. The locations of

Table IV.   Length parameter *l* for center crack panel subjected to uniform stress and concentrated load.

Uniform stress		Concentrated Load	
a (cm)	l (cm)	a (cm)	l (cm)
2.54	10.16	2.54	7.32
5.08	7.62	5.08	5.23
7.62	3.74	7.62	2.99
10.16	1.49	10.16	1.64

$[(dW/dV)^{\text{max}}_{\text{min}}]_e$ and $[(dW/dV)_{\text{min}}]_g$ correspond to the points *L* and *G*, respectively, in Figures 1.19(a) and 1.19(b). The point *L* always coincides with the crack tip while *G* can move depending on the specimen dimension and loading type. The data in Table IV for the case of uniform stress loading show that the conditions $a + l = d$ is satisfied when the initial half crack length *a* is less than or equal to 5.08 cm. This corresponds to the point *G* lying on the specimen boundary. For longer crack length, *l* becomes smaller than the ligament distance $d-a$ and the point *G* moves to the interior of the specimen. Since both points *L* and *G* are sites dominated by dilatation in contrast to distortion, they tend to fail by fracture. Therefore, *l* may be regarded as a measure of the stability of a system. Referring to Table IV, the cases corresponding to $l = 10.16$ and 7.62 cm imply that fracture occurs across the entire ligament $d-a$ as σ reaches its critical value. For the remaining two cases of $l = 3.74$ and 1.49 cm, however, fracture is more localized as the distance between L and G is now reduced. It is difficult to recognize this stabilizing tendency experimentally because all specimens are subject to overload in reality which is difficult to assess quantitatively as the specimen stiffness changes with the initial crack length. This effect is demonstrated even more vividly for the case of concentrated load in Figure 1.19(b). Now that the load is more localized, l or subcritical material damage decreases accordingly. The point G always lies to the interior of the specimen. The crack behavior under the loading type in Figure 1.19(b) is concluded to be more stable than that in Figure 1.19(a).

The effect of material nonhomogeneity can also be reflected through l [32]. Consider the situation of an elliptically shaped inclusion with modulus E_a embedded in a matrix material with modulus E_m as shown in Figure 1.20. For $b/a = 0.5$, a plot of the normalized length $l/(d-a)$ versus E_a/E_m is given in Figure 1.21. All the curves increase in magnitude as E_a tends to E_m and approach the limit $l = d-a$ depending on c/d. The value $l/(d-a) = 1$ corresponds to having G moved to the specimen boundary. As the system homogeneity is increased, the energy density distribution along

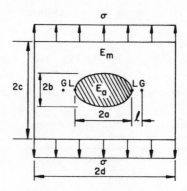

Figure 1.20. Inclusion in matrix.

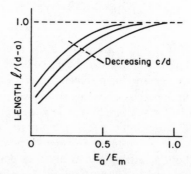

Figure 1.21. Influence of material inhomogeneity on l.

the path of expected failure becomes more uniform. This increases the instability of the system because fracture would then occur more suddenly through the specimen. The failure becomes more localized as G moves closer to L when E_a/E_m deviates from unity. Localized inhomogeneity enhances local failure, a phenomenon that is typical of any multi-phase material. The same conclusion applies to concrete where $E_a/E_m > 1$.

A quantitative assessment of the degree of inhomogeneity of a multi-phase system can thus be made by comparing l with the specimen or structural component dimension. For composite materials such as concrete, it is beneficial to have l small so that damage can be kept at a subcritical level and the chance of catastrophic failure is minimized. The stress and failure analysis of inhomogeneous systems need not be overly complicated since a body could be divided into many subregions and only those which undergo active interaction with load require more detailed attention. Damage zones can be defined such that solution accuracy is not required everywhere. Such concept has already been in practice for a long time in analyzing crack problems where accuracy is needed only in the region close to the crack singular point. The real value of l is that it provides a means for optimizing the load, geometric and material parameters in terms of the system instability to failure.

1.8 General discussion

As modern technology is expanding and advancing at a rapid rate, the need for better control of material and/or structure failure becomes more demanding. From the economical viewpoint alone, concrete material is attractive. Concrete is being used more and more in floating marine structures, storage tanks, etc., aside from its more conventional application in building, road and bridge construction. Because of this demand, much remains to be done in further understanding of concrete behavior under service loading conditions and environments.

Clearly, the conventional approach of collecting more test data will serve little or no useful purpose. Much better characterization of test specimens and sophistication in the analysis of data are needed. To be always kept in mind is how and where the test data could be used to improve structure design. With a limited knowledge of assessing concrete strength, more emphases have to be placed on the interaction between load type and material inhomogeneity. The dramatic contrast between the tensile and compressive strength of concrete is indicative of the dominant role that material inhomogeneity plays. Structures, by nature, must sustain a variety of mixed load conditions such as creep, fatigue and occasionally impact. Unless the material's resistance to failure is identified with load type, no useful information can be gained.

While the needs point towards addressing the admittedly difficult task of numerically and/or experimentally resolving the complex behavior of concrete, it is essential to first establish a sound base of approach upon which knowledge can be accumulated. More sophisticated analyses should be adopted to provide guidelines on the type and degree of characterization that is required to adequately carry out the experimental and analytical study of a given task. Characterization simply for characterization's sake tends to obscure the main objective of the task which is to translate test data to design. The long-range need is for the development of models to take into account the distribution of the shape, type and location of aggregates in relation to loading type and to determine the conditions under which the influence of the secondary inhomogeneities such as pores and microcracks in the cement also become important. Because inhomogeneity due to material properties and damage modes are the primary concern in composites, emphasis should be placed on recognizing the critical parameters that govern concrete behavior. The most significant contribution therefore lies in establishing the relation between local material damage and global specimen response.

References

[1]   Mindess, S., The cracking and fracture of concrete: an annotated bibliography, 1928–1980, *Materials Research Series,* Report No. 2 I.S.S.N. 0228-4251, The University of British Columbia, Vancouver (1981).

[2]   Sih, G.C., Hilton, P.D., Badaliance, R., Shenberger, P.S. and Villarreal, G., Fracture mechanics of fibrous composites, *ASTM Special Technical Publication* No. 521, pp. 98–132 (1973).

[3]   Popovics, S., Fracture mechanism in concrete: How much do we know?, *Journal of Engineering Mechanics Division,* ASCE, Vol. 95, EM3, pp. 531–544 (1969).

[4]   Sih, G.C., Fracture toughness concept, *ASTM Special Technical Publication* No. 605, pp. 3–15 (1976).

[5]   Plane strain crack toughness testing of high strength metallic materials, edited by W.F. Brown, Jr. and J.E. Srawley, *ASTM Special Technical Publication* No. 410 (1966).

[6]   Carpinteri, A. and Sih, G.C., Damage accumulation and crack growth in bilinear materials with softening: application of strain energy density theory, *Institute of Fracture and Solid Mechanics Technical Report,* IFSM-83-115, (1983).

[7]   Sih, G.C., Mechanics of crack growth: geometrical size effect in fracture, *Fracture Mechanics in Engineering Application,* edited by G.C. Sih and S.R. Valluri, Sijthoff and Noordhoff, pp. 3–29 (1979).

[8]   Sih, G.C., The mechanics aspects of ductile fracture, *Continuum Models of Discrete Systems,* edited by J.W.Provan, University of Waterloo Press, pp. 361–386 (1977).

[9]   Oldroyd, J.G., The effect of small viscous inclusions on the mechanical properties of an elastic solid, Grammel, editor, Springer Verlag Berlin, pp. 304–313 (1956).

[10] Mackenzie, J.K., The elastic constants of a solid containing spherical holes, *Proc. Phys. Soc. London,* Series B, Vol. 63, pp. 2–11 (1950).

[11] McNeeley and Lash, S.D., Tensile strength of concrete, *Proc. American Concrete Institute,* Vol. 60, pp. 751–761 (1963).

[12] Alexander, K.M., Strength of cement-aggregate bond, *Proc. American Concrete Institute,* Vol. 56, pp. 377–390 (1960).

[13] Hsu, T.T.C. and Slate, F.O., Tensile bond strength between aggregate and cement paste or mortar, *Proc. American Concrete Institute,* Vol. 60, pp. 465–486 (1963).

[14] Sih, G.C., A special theory of crack propagation, *Methods of Analysis and Solutions of Crack Problems,* edited by G.C. Sih, Noordhoff International Publishing, Leyden, pp. 21–45 (1973).

[15] Sih, G.C., A three-dimensional strain energy density factor theory of crack propagation, *Three Dimensional Crack Problems,* edited by M.K. Kassir and G.C. Sih, Noordhoff International Publishing, Leyden, pp. 15–53 (1975).

[16] Newman, K. and Newman, J.B., Failure theories and design criteria for plain concrete, *Structure, Solid Mechanics and Engineering Design,* edited by M. Te'eni, Proceeding of the Southampton 1969 Civil Engineering Materials Conference, Wiley Interscience, pp. 963–995 (1969).

[17] Sih, G.C., Chen, E.P., Huang, S.L. and McQuillen, E.J., Material characterization on the fracture of filament-reinforced composites, *Journal of Composite Materials,* Vol. 6, pp. 167–185 (1975).

[18] Sih, G.C., Fracture mechanics of adhesive joints, *Journal of Polymer Engineering and Science,* Vol. 20, No. 14, pp. 977–981 (1980).

[19] Gillemot, L.F., Criterion of crack initiation and spreading, *International Journal of Engineering Fracture Mechanics,* Vol. 8, pp. 239–253 (1976).

[20] Ivanova, V.S., Maslov, L.I. and Burba, V.I., R-similarity criterion of plastic deformation instability and its use for K_{1c} determination for steels, *Journal of Theoretical and Applied Fracture Mechanics, Vol. 2, No. 3* (forthcoming).

[21] Sih, G.C. and Macdonald, B., Fracture mechanics applied to engineering problems – strain energy density fracture criterion, *Engineering Fracture Mechanics,* 6, pp. 361–386 (1974).

[22] Naus, D.J. and Lott, J.L., Fracture toughness of Portland cement concretes, *Journal of the American Concrete Institute,* Vol. 66, No. 6, pp. 481–489 (1969).

[23] Walsh, P.F., Crack initiation in plain concrete, *Magazine of Concrete Research,* Vol. 28, pp. 37–41 (1976).

[24] Hillemeier, B. and Hilsdorf, H.K., Fracture mechanics studies on concrete compounds, *Cement and Concrete Research,* Vol. 7, pp. 523–536 (1977).

[25] Zaitsev, J.W. and Wittmann, F.H., Crack propagation in a two-phase material such as concrete, *Proceedings of the 4th International Conference on Fracture,* Waterloo, Canada (1977).

[26] Swamy, R.N., Influence of slow crack growth on the fracture resistance of fibre cement composites, *International Journal of Cement Composites,* Vol. 2, No. 1, pp. 43–53 (1980).

[27] Petersson, P.E., Crack growth and development of fracture zones in plain concrete and similar materials, Report TVBM-1006, Lund Institute of Technology, Division of Building Materials (1981).

[28] Sih, G.C. and Matic, P., A pseudo-linear analysis of yielding and crack growth: strain energy density criterion, *Defects, Fracture and Fatigue,* edited by G.C. Sih and J.W. Provan Martinus Nijhoff Publishers, The Hague, pp. 223–232 (1983).

[29] Hilton, P.D., Gifford, L.N. and Lomacky, O., Finite element fracture mechanics of two dimensional and axisymmetric elastic and elastic-plastic cracked structures, *Naval Ship Research and Development Center,* Report No. 4493 (1975).

[30] Carpinteri, A., Size effect in fracture toughness testing: a dimensional analysis approach, *Analytical and Experimental Fracture Mechanics,* edited by G.C. Sih and M. Mirabile, Sijthoff and Noordhoff, pp. 785 –797 (1981).

[31] Carpinteri, A., Static and energetic fracture parameters for rocks and concretes, *Materials and Structures,* 14, pp. 151–162 (1981).

[32] Chu, R.C., Local and global stationary values of strain energy density in a nonhomogeneous medium, Department of Mechanical Engineering and Mechanics, Ph.D. Thesis, Lehigh University (1983).

[33] Sih, G.C. and Madenci, E., Crack growth resistance characterized by the strain energy density function, *International Journal of Fracture Mechanics,* in press (1983).

Evaluation of concrete fracture

2.1 Introduction

It is well-known that concrete behavior is highly influenced by cracks at the mortar-aggregate interface and defects in the mortar matrix. These defects or cracks can spread as the load is increased. The means for detecting the initial imperfections are essential for the development of analytical models. Discussed are several of the non-destructive testing techniques for finding the size and location of flaws. Different loading types such as compression, tension and biaxial compression and tension are also known to drastically affect the load transfer characteristics in concrete. Damage by cracking can be analyzed by application of Linear Elastic Fracture Mechanics that utilizes the concept of stress intensity factors. In problems where Mode I and II crack extension prevail simultaneously, the maximum normal stress and strain energy density criterion may be used to determine the direction of crack initiation. A modi-fied version of the maximum stress criterion is presented for the case of an interface crack.

2.2 Mechanism of failure in concrete

The mechanism of failure in concrete and the particular nature of the stress-strain relationships for concrete under uniaxial, biaxial and triaxial loading have been studied experimentally by many investigators over the last few years.

The mechanism of progressive damage in concrete has been detected by means of different experimental techniques.

Technical observations of damage.

a) X-rays observation.

This technique has been extensively used by Robinson [1] and Slate and Olsefski [2]. X-rays exposures in several directions of a certain volume of concrete under loading allow the detection of micro-cracks at different levels of loading.

b) Optical observation.

This technique has been used by Dhir and Sangha [3]. Concrete cylinders were loaded in compression at different levels and longitudinal slices of the specimens were coated with fluid ink. The ink tracks delineated the discontinuities on optical observation by microscope. Some maps of microcracks were drawn.

c) Ultrasonic-wave speed detection.

This method can be utilized to describe the 'density' of microcracks at various levels of loading in connection with ultrasonic wave speed in the material [4]. A measure of the anisotropic change of concrete can be detected by using more sensors.

d) Acoustic emission.

It is well-known that any damage in concrete (debonding, matrix cracking, etc.) causes noise. By means of an appropriate technique, the 'signals' of acoustic events are analyzed and connected with concrete damage at various levels of loading [5].

Mechanism and phases of damage. Thanks to the above-mentioned techniques, it was possible to observe that microcracks existing in unloaded concrete are more frequently located in a high porosity layer of the transition zone (interface) between the coarsest aggregates and mortar (Figure 2.1).

It has been established by now that there are two main mechanisms of damage in concrete under loading [6]:

(a) debonding at the interface between aggregate and cement paste;
(b) microcracking in the matrix.

The following phases of damage occur under the loading process (Figure 2.2):

(a) the microcracks at the interface start to progress as 'debonding' cracks;
(b) at higher loading, the microcracks at the interface start to branch inside the mortar;

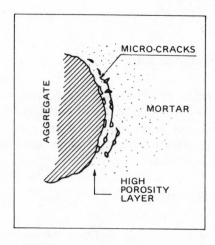

Figure 2.1   Aggregate-mortar bonding

INCREASE OF THE STRESS FIELD

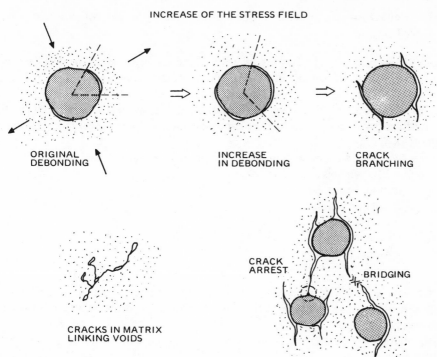

ORIGINAL
DEBONDING

INCREASE
IN DEBONDING

CRACK
BRANCHING

CRACK
ARREST

BRIDGING

CRACKS IN MATRIX
LINKING VOIDS

Figure 2.2   Mechanism of damage

(c) cracks in the mortar tend to progress orthogonally to principal strains and new cracks are formed in mortar linking voids;

(d) 'bridging' between different cracks tends to cause the 'failure' of the material, while the aggregate works as a 'crack arrestor'.

The exceptional mechanical characteristics of some high-strength concretes (auto-claved concretes) are mainly due to the past aggregate high quality bond [7].

Physical properties of constituents and damage. The number, location and extent of pre-existing cracks at the interface depend mainly on:

(a) type of cement;
(b) mineralogical nature of aggregate;
(c) geometry of aggregate;
(d) water/cement ratio;
(e) curing conditions.

The evolution of pre-existing cracks under loading depends mainly on:

(a) aggregate/matrix stiffness ratio;
(b) type of matrix-aggregate bond;
(c) percentage of voids in the matrix.

2.3   Behaviour of concrete under loading related to the mechanism of failure

Uniaxial compression. The process by which concrete deforms in an uniaxial compression test and ultimately fails by development and coalescence of microcracks has been well-documented by many investigators.

The importance of recording the evolution of the volumetric strain, $\Delta V/V$, in conjunction with the stress-strain relationship has also been underlined by many authors [8–11, 14].

The curves for a concrete cylinder in compression test are plotted qualitatively in Figure 2.3. The symbols stand for:

σ compressive stress
ϵ_L longitudinal strain
ϵ_T transversal strain
$\Delta V/V$ volumetric strain ($\epsilon_L + 2\epsilon_T$; $\epsilon_L < 0$; $\epsilon_T > 0$)
ν Poisson ratio ($|\epsilon_T : \epsilon_L|$)

Some remarks should be made:

(i) above a certain stress σ_i the curve of volumetric strain has an inflection point, F; a sort of 'swelling' is active at this stress level parallel to the common decreasing volume phenomenon;

(ii) above a critical stress σ_{cr}, the volume starts to increase rather than continuing to decrease.

These results are obtained by measuring the strain referred to a length being at least three times as large as the maximum aggregate diameter; in this context, *cracks are not evaluated as material discontinuities.*

Referring to Figure 2.3, five intervals of stresses can be identified with points on the $\sigma - \Delta V/V$ ascending branch (phenomenological aspects):

(O–E')	I	settlement;
(E'–E)	II	linearity;
(E–F)	III	increasing deformability;
(F–M)	IV	increasing deformability + swelling;
(M–L)	V	increasing deformability + non-stable swelling.

In each interval I → V the predominantly active mechanism of damage has been identified by means of technical observations (see also Figure 2.3):

(O–E')	I	cracking and pore closure;
(E–E)	II	crack opening with absence of new internal free surfaces;
(E–F)	III	debonding predominates and is the main factor responsible for non linearity (new internal free surfaces at the interface aggregate-matrix);
(F–M)	IV	bond cracks begin to deviate into the matrix (new internal surfaces in the matrix);
(M–L)	V	cracks become non-stable in the matrix (new internal surfaces in the matrix);
(L–)	VI	coalescence of matrix cracks dominates (beyond the ultimate stress).

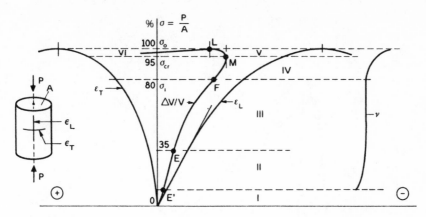

Figure 2.3   Monoaxial compression and deformation parameters

It has been found that, in most cases, sufficiently closed matrix cracks coalesce as a result of interaction in the stress field. An inclined failure surface can be formed which passes through the entire specimen [11,12].

When a continuous crack pattern has developed extensively and coalescence is active, the load carrying capacity of the concrete decreases and the stress-strain curve begins to descend (*softening*). In a cement paste specimen, cracks extend through the specimen in a straight line; for mortar and normal concrete, however, the crack pattern has a 'meandering' path and tends to go around the aggregate instead of going through it. This energy consuming process stabilizes the crack growth.

The descending branch of the stress-strain curve depends on the 'strain rate'. Experimental stress-strain curves for different strain rates are represented in Figure 2.4. It should be noted that in reinforced concrete specimens the presence of stirrups makes the reduction of stress in the descending branch of the curve slower (Figure 2.5). A procedure for determining the descending branch consists in recording a cyclic program, as shown in Figure 2.6. The envelope can approximately be considered as the effective descending branch for the material; the strain rate characterizing the particular envelope curve can be related to the time needed for every loop.

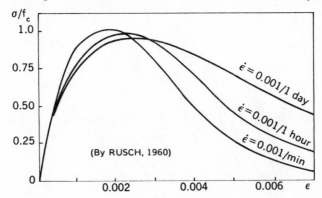

Figure 2.4   Experimental stress-strain curves in compression for various strain rates $\dot{\epsilon}$.

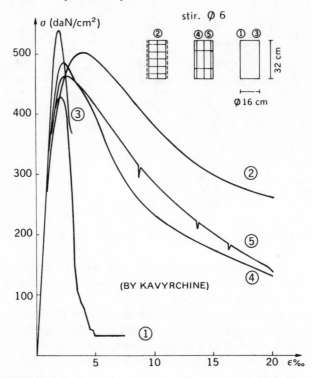

Figure 2.5   Experimental stress-strain curves in compression in cylindrical specimens with stirrups.

The aspect of the volumetric strain under cyclic loading (loops) is plotted in Figure 2.7. The type of hysteretic loops [11,14] is worthy of remark: they are clock-wise before point M ($\sigma = \sigma_{cr}$), as shown in Figure 2.3, and in an opposite sense after this point. This is due to the non-stable crack pattern after point *M*.

The stress-strain relationship between 'constant levels' under repeated loading is shown in Figure 2.8 [13]. In many cases the increasing load curve shows a slight

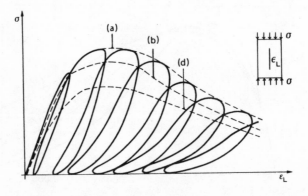

Figure 2.6   Repeated loads in compression and strain loops.

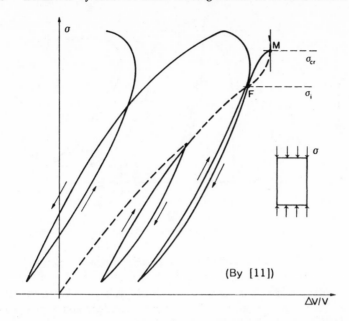

Figure 2.7   Repeated loads in compression and volumetric strain loops.

change in concavity. At the beginning, the loading branch presents some concavity toward the strain axis; after a certain number of repeated loadings, it becomes linear at first and then convex towards the same axis. As a general rule, the unloading branch is always convex towards the strain axis.

The Poisson's ratio during the cyclic loading between 'constant levels' of stress is recorded in Figure 2.9 [14]. Very high values (~ 1.5) are reached* after six cycles, the maximum load being very close to the ultimate strength.

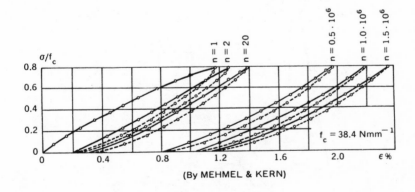

Figure 2.8   Repeated loads in compression.

*Note that cracks are not here evaluated as material discontinuities; in this sense, the voids are included in the strain computation and the high values of ν can be justified. This confirms the above-mentioned 'swelling'.

37

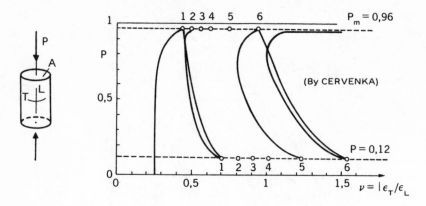

Figure 2.9   Repeated load-Poisson's ratio curve in compression.

Uniaxial tension. Very few investigations regarding the stable stress-strain curve for concrete in uniaxial tension are reported in the literature. In fact, a very stiff testing machine is necessary for this kind of test. Such tests have been carried out more recently by Evans and Marathe [15], Heilman, Hilsdorf and Finsterwalder [16], Hillerborg [17], Terrien [18] and Di Leo [19].

When recording the stress-strain curve, the technique of measuring the strain is relevant. In fact, two possibilities are found in the literature, either measuring the change in length of the whole specimen or measuring the change in length in small zones [17]. Obviously, the results are different when considering the local micro-cracking in small zones of the specimen at maximum load.

The type of diagram evaluated by measuring the change in length of the whole specimen is shown in Figure 2.10. Some remarks should be made about this diagram:

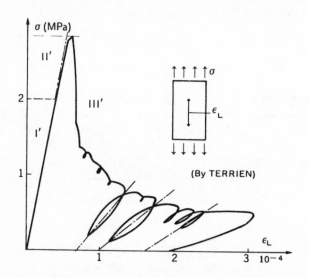

Figure 2.10   Uniaxial tensile stress-strain curve.

(i) in the range before the peak the stress-strain curve slightly deviates from the linear behaviour; there is a first stage (I') in which the elastic behaviour can be considered as a consistent hypothesis;

(ii) in a second stage (II'), after the elastic stage (I') and before the peak, some small cracks (mainly debonding cracks) appear in a small zone of the specimen. These cracks are stable and few in number. The stress-strain curve shows a small deviation from linearity until the peak is reached;

(iii) in a third stage (III'), a first development of microcracks occurs, especially in the small zone in which damage was previously in progress. This fact involves a sharp fall in the tension force carrying capacity and can only be followed by means of a very stiff servo-displacement machine or by special devices consisting of stiff bars in parallel with the concrete specimen [16, 19].

In any case, the maximum strain in tension can be over four times as high as the strain corresponding to the peak load.

Biaxial compression and/or tension. As far as the biaxial state of stress is concerned, the experimental work by Kupfer, Hilsdorf and Rüsch [20] is the main reference. By means of a particular 'brush' device, it was possible to obtain a homogeneous state of stress in the specimen. A constant principal tension ratio σ_1/σ_2 (with $\sigma_3 = 0$) was realized until failure. These experiments show a volumetric strain behaviour, as far as biaxial stresses are concerned, which is similar to that evaluated for uniaxial compression (Figure 2.11). In this case, a locus for each point E,F,M,L can be represented in the σ_1, σ_2 plane. Such loci are closed curves around the origin (Figure 2.12a) and may be named 'discontinuity curves', while the locus (f) may be named 'fracture locus'. The major distance between the curves in the compression-compression range is due to the influence of 'friction' on the failure phenomenon.

A qualitative description of the cracking approaching the situation represented by ultimate state (curve *l*) is shown in Figure 2.12b. The anisotropy of the material in the range between the curves (*m*) and (*l*) is evident. An investigation to differentiate

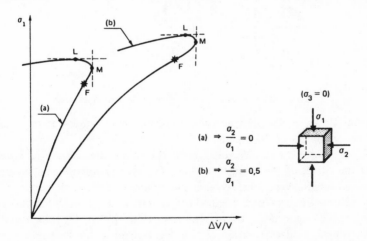

Figure 2.11   Volumetric strain in monoaxial and biaxial compression.

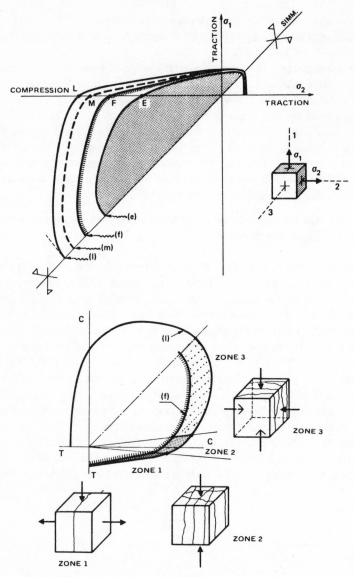

Figure 12a,b   Discontinuity curves in biaxial stress state.

the different states of the material can be conducted by means of technical observations (see section 2.1 of this chapter).

The discontinuity curves are sensitive to the *strain rate*. Figure 2.13 shows the change of the curves of the volumetric strain, for uniaxial compression, in connection with different loading velocities. As far as the influence of the strain rate is concerned, it may be affirmed that an increase in the strain rate makes concrete 'more brittle', in the sense that the discontinuity curves tend to be closer $[(e) \rightarrow (l)]$ to one another.

A phenomenological explination of this event could be the influence of viscosity with regard to the compact and microcracked material.

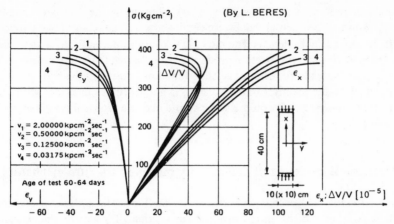

Figure 2.13   Variation of the longitudinal and transversal deformation as well as volume change at various loading rates in compression.

Regarding the *aggregate size,* it has been established that a decrease in the maximum aggregate size makes concrete 'more brittle', in the sense that the discontinuity curves tend again to be closer $[(e) \rightarrow (l)]$ to one another. This is due to the path of micro-cracks which, for concretes with small aggregate, need less tortuosity (and then less energy demand) to coalesce.

Very little information is currently available on the 'softening' of concrete in biaxial state of stress; experimental programs in this sense could be undertaken to evaluate this phenomenon in detail.

A possible representation of the limit curves, namely ultimate curve (*l*) and fracture curve (*f*), is shown in Figure 2.14. It is in agreement with the experimental data [20]

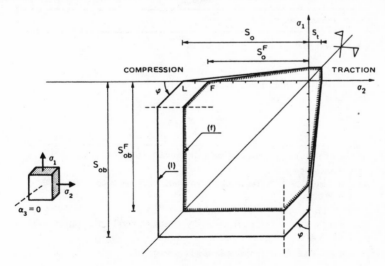

Figure 2.14   Biaxial stress domain

41

and contains four parameters $(\alpha, \beta, \gamma, \delta)$ in addition to the ultimate compressive stress S_c. Some experimental values concerning the parameters obtained by several authors are shown in Table 2.1. The curve (f) is very important also with respect to cyclic loading. In fact, for stresses inside the closed curve (f), the permanent deformation is limited, while for stresses outside the closed curve (f) the permanent deformation, which is due mainly to micro-cracks, is noteworthy. Consistent with this last consideration, the curve (f) could also be considered as a *fatigue limit* curve in biaxial stress state.

2.4 Mechanical modeling of concrete by means of fracture mechanics

Two different approaches are currently developed in order to characterize the cumulative effects of progressive micro-cracking of concrete under sustained loads:

(i) *Mathematical Relationships* between stress and strain by using theoretical concepts (i.e. plasticity, damage, etc.) or experimental correlations.

(ii) *Mechanical Models* with one or several phases, including different kinds of cracks, pointing out local phenomena (debonding, crack propagation and/or crack arrest) and their consequences for global behaviour of the material.

In the second approach, the concrete volume element is represented by means of a generally cubic cell, including one or more phases with initial flaws prone to grow under loading.

TABLE 1

EXPERIMENTAL VALUES

$$\alpha = \frac{S_t}{S_o} \cong 0.10 \qquad\qquad \varphi = \sim \frac{\pi}{4}^{[2][3]}$$

$$\beta = \frac{S_{ob}}{S_o} = 1.16^{[1][2]}; \ 1.20^{[3]}; \ 1.25^{[4]}$$

$$\gamma = \frac{S_o^F}{S_o} = 0.83^{[5]} \qquad\qquad \delta = \frac{S_{ob}^F}{S_o} \cong 1.00^{[4]}$$

(1)	J. WASTIELS	(Vrije Universiteit, Brussel) IABSE Colloquium 1979
(2)	H.B. KUPFER H.K. HILSDORF H. RÜSCH	(Technische Hochschule, München) A.C.I. Journ. 1969
(3)	C.Y. LIU A.H. NILSON F.O. SLATE	(Cornell University, Ithaca, N. York) ASCE Proceed. 1972
(4)	G.W.D. VILE	(Imperial College, London) 1968
(5)	A. DI LEO A. DI TOMMASO R. MERLARI	(Università di Bologna) 1978

Behaviour of a main crack in a biaxial stress field of an elastic isotropic material. Considering a homogeneous two-dimensional stress field with a line* embedded orthogonally to the plane of the stress, σ_β is the normal tension at infinity producing the *'opening effect'*, and τ_β is the shear at infinity producing the *'sliding'* effect on the main crack and σ_1 and σ_2 are the principal stresses (Fig. 2.15a).

The stress field close to the crack tip ($r \ll a$) is represented in polar coordinates by the following expressions (for notations see Figure 2.16a, in which the positive stresses are represented).

$$\sqrt{2\pi r} \begin{bmatrix} \sigma_r \\ \hline \sigma_\vartheta \\ \hline \tau_{r\vartheta} \end{bmatrix} = \cos\frac{\vartheta}{2} \left[\begin{array}{c|c} 1 + \sin^2 \vartheta/2 & \frac{3}{2}\sin\vartheta - 2\tan\vartheta/2 \\ \hline \cos^2\vartheta & -\frac{3}{2}\sin\vartheta \\ \hline \sin\vartheta & 3\cos\vartheta - 1 \end{array} \right] \begin{bmatrix} K_I \\ K_{II} \end{bmatrix} \qquad (2.1)$$

where:

$$K_I = \sigma_\beta \sqrt{\pi a}; \qquad K_{II} = \tau_\beta \sqrt{\pi a} \qquad (2.2)$$

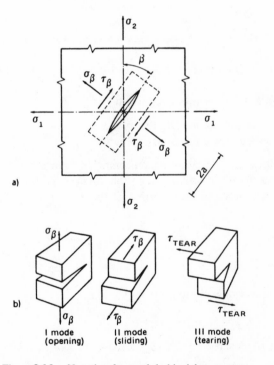

Figure 2.15 Notation for crack in biaxial stress state.

*The fundamental concepts of fracture mechanics are considered known to the reader. This paragraph is devoted to the introduction of notation and to the recall of the main topics to be developed in the next part.

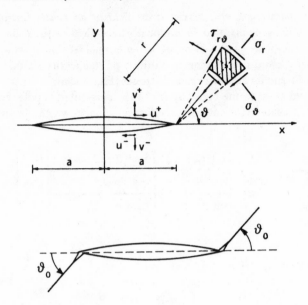

Figure 2.16   Notation for crack tip stress and branching.

are the *stress intensity factors.** Note the *'singularity'* at the tip of the crack (for $r \to 0$).

Two fracture 'criteria' are now considered:

(i) Maximum Stress Criterion (K_c-theory);
(ii) Strain Energy Density Criterion (S_c-theory).

(i) According to the K_c-theory, crack branching starts from the 'tip' in a radial direction [21] in the plane orthogonal to the direction of maximum hoop stress σ_ϑ; propagation corresponds to a 'critical value' of the stress intensity factor, $K_I = K_{IC}$:

$$\begin{cases} \left(\dfrac{\partial \sigma_\vartheta}{\partial \vartheta}\right)_{\vartheta = \vartheta_0} = 0 \\ \sigma_{\vartheta_0} \sqrt{2\pi r} = K_{IC} \end{cases} \tag{2.3a}$$

or, (excluding the obvious solution $\vartheta = \pm \pi$) [21]:

$$\begin{cases} K_I \sin \vartheta_0 + K_{II} (3 \cos \vartheta_0 - 1) = 0 \\ \cos \vartheta_0/2 \left[K_I \cos^2 \vartheta_0/2 - 3/2 (K_{II} \sin \vartheta_0)\right] = K_{IC} \end{cases} \tag{2.3b}$$

*The expressions (2.1) of the stress components σ_r, σ_ϑ, $\tau_{r\vartheta}$ in polar coordinate are valid for both 'plane strain' and 'plane stress' conditions. As is well-known, when the axis perpendicular to the $x-y$ plane (Fig. 2.16a) is denoted by z, the plane strain condition can be expressed as $\epsilon_z = \partial w/\partial z = 0$ and $\sigma_z = \nu(\sigma_x + \sigma_y)$, where w is the displacement component in the z direction, ϵ_z denotes the strain in the same direction and $\sigma_x, \sigma_y, \sigma_z$ are the stress components. In the 'plane stress' condition we have $\sigma_z = 0$.

The expressions of the displacement components u, v around the crack tip in the x and y directions respectively, are different for 'plane strain' and 'plane stress. conditions respectively.

where ϑ_0 is the angle of crack branching (see Figure 2.16b). Equation (2.3a), or (2.3b), is the parametric equation of the *limit curve*, in the $K_I - K_{II}$ geometrical plane, for crack branching.**

(ii) According to the S_c-theory proposed by G.C. Sih [22], one considers the Strain Energy Density Factor:

$$S = a_{11}K_I^2 + 2a_{12}K_IK_{II} + a_{22}K_{II}^2 \tag{2.4}$$

with the coefficients valid for the plane strain problem:

$$a_{11} = \frac{1}{16\mu}\left[(3 - 4\nu - \cos\vartheta)(1 + \cos\vartheta)\right]$$

$$a_{12} = \frac{1}{16\mu} 2\sin\vartheta \left[\cos\vartheta - (1 - 2\nu)\right]$$

$$a_{22} = \frac{1}{16\mu}\left[4(1 - \nu)(1 - \cos\vartheta) + (1 + \cos\vartheta)(3\cos\vartheta - 1)\right]$$

where μ is the tangential modulus of elasticity and ν the Poisson's ratio.

The crack extension condition, according to the S_c-theory, is:

$$\begin{cases} \left(\dfrac{\partial S}{\partial \vartheta}\right)_{\vartheta = \vartheta_0} = 0 \\[2mm] S = S_{cr} = \dfrac{K_{IC}^2(1 - 2\nu)}{4\mu} \end{cases} \tag{2.5}$$

The relations (2.5) state that the crack branching angle ϑ_0 (Fig. 2.16b) corresponds to a stationary value (minimum) of S and propagation occurs for a critical value of $S = S_{cr}$ depending on the material. Equation (2.5) is the parametric equation of the *limit curve* in the plane K_I, K_{II}.

By introducing the *non-dimensional stress intensity factors*:

$$K_I^* = \frac{K_I}{K_{IC}}; \qquad K_{II}^* = \frac{K_{II}}{K_{IC}} \tag{2.6}$$

it is possible [23] to plot the *limit curves* corresponding to the K_c-theory and S_c-theory from (2.3) and (2.5) in the geometrical plane of the non-dimensional stress intensity factors K_I^*, K_{II}^*, as shown in Figure 2.17. The diagram is valid for every elastic material. The process of the friction doesn't allow one to extend the diagram in the range $\sigma_\beta < 0$ (this should lead to a 'fictitious' $K_I^* < 0$). In fact, when $\sigma_\beta < 0$, friction arises on the crack free surfaces and the singularity at its tip disappears, that is the 'effective' K_I^* is null. In this last case, it follows that the crack extension

---

**The stationary and critical conditions expressed by equations (2.3a), are valid for 'plane stress' as well as for 'plane strain' conditions, when an appropriate value of the fracture toughness K_C is considered. In this sense, K_C is a function of the thickness B of the stressed material, i.e. $K_C = K_C(B)$. When B tends to infinity, K_C tends to K_{IC}. For small and intermediate values of B, the corresponding higher values of fracture toughness are designated by K_C. The plane stress toughness K_C and the plane strain toughness K_{IC} have been widely treated in literature, but systematic data for various materials are still limited.

conforms to mode II ($K_I = 0$, $K_{II} \neq 0$). Then, for $\sigma_\beta < 0$ (compressive normal tension on crack face) the critical condition, in the extended K_c-theory, is [24]:

$$|K_{II}| - f|K_I| = K_{IIC} \tag{2.7}$$

where K_{IIC} is the mode II critical stress intensity factor, correlated to K_{IC} in each theory (see the intersection of the curves with K_{II}^* axis in Figure 2.17); $K_I = \sigma_\beta \sqrt{\pi a}$; $K_{II} = \tau_\beta \sqrt{\pi a}$ and f is the friction coefficient betwen the crack surfaces in contact.

The condition (2.7) allows the description of the limit curve in the geometrical plane K_I^*, K_{II}^*, also in the range $\sigma_\beta < 0$ (i.e. fictitious $K_I^* < 0$), as shown in Figure 2.18.

In the case of a homogeneous state of stress at infinity with principal stresses σ_1, σ_2 and 'line-crack' with length $2a$ inclined to the angle β with respect to direction σ_2, it follows that (Fig. 2.15):

$$K_I = K_I(\beta) = \sigma_\beta \sqrt{\pi a} = \left(\frac{\sigma_1 + \sigma_2}{2} - \frac{\sigma_2 - \sigma_1}{2} \cos 2\beta \right) \sqrt{\pi a}$$

$$\tag{2.8}$$

$$K_{II} = K_{II}(\beta) = \tau_\beta \sqrt{\pi a} = \left(\frac{\sigma_2 - \sigma_1}{2} \sin 2\beta \right) \sqrt{\pi a}$$

With the normalizations:

$$\sigma_1^* = \frac{\sigma_1 \sqrt{\pi a}}{K_{IC}} \qquad \sigma_2^* = \frac{\sigma_2 \sqrt{\pi a}}{K_{IC}} \tag{2.9}$$

and recalling equation (2.6), one can re-write equation (2.8) as follows:

$$\sigma_1^* = K_I^* - \frac{K_{II}^*}{\sin 2\beta} (1 - \cos 2\beta)$$

$$\tag{2.10}$$

$$\sigma_2^* = K_I^* + \frac{K_{II}^*}{\sin 2\beta} (1 + \cos 2\beta)$$

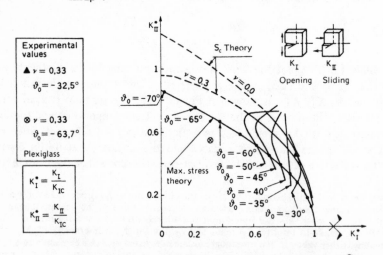

Figure 2.17   Limit curves in $K_I^* - K_{II}^*$ geometrical plane.

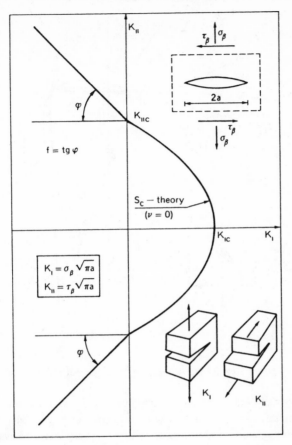

Figure 2.18   Limit curve in $K_I - K_{II}$ geometrical plane with frictional effect.

After evaluating all the couples of values K_I^*, K_{II}^* corresponding to the crack extension (as plotted in Figure 2.17), one obtains a limit curve for each main crack inclination ($\beta = \bar{\beta} =$ const.), in the σ_1^*, σ_2^* geometrical plane. In this case also, this is valid for $\sigma_\beta \geqslant 0$. This condition, taking into account (2.6) and (2.8), gives:

$$\sigma_1^* \geqslant \sigma_2^* (\cos 2\beta - 1)/(\cos 2\beta + 1) \tag{2.11}$$

Equation (2.11) is the validity condition for each limit curve $\beta = \bar{\beta}$ in absence of friction. The introduction of friction allows the representation of Figure 2.19, in which a friction coefficient is assumed.

The approach outlined above does not take into account the normal stress σ_p 'collinear' to the crack in the propagation criteria. In fact, if one considers the stress field close to the crack tip, the first term of series representing the stresses is inversely proportional to the square root of the distance r from the crack tip (singular term) — see equation (1) — while the second term is independent of such distance (non singular term). Therefore, the closer to the crack tip the more negligible the second term becomes in comparison with the first one.

47

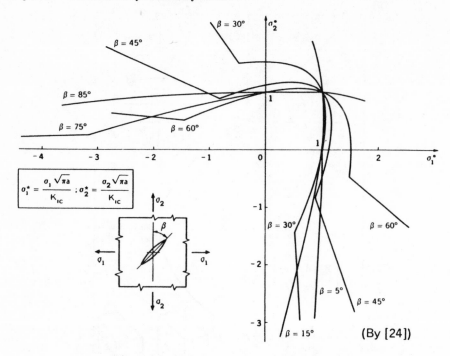

Figure 2.19   Limit curves in $\sigma_1^* - \sigma_2^*$ geometrical plane.

Such an approximation, as recently pointed out [25], is not generally satisfactory. A loading parameter $\lambda = \sigma_p/\sigma_\beta$ and a material parameter r_0 (distance from the crack tip where the propagation condition is established) are necessary to evaluate the crack propagation conditions. This fact modifies the limit curve represented in Figure 2.17 in the sense shown in Figure 2.20, according to the analysis worked out in [26].

In the following, the influence of collinear stress will not be taken into account. In fact, the effort of the present approach consists in establishing "very simple" concrete mechanical models with a minimum of mechanical and loading parameters.

Behaviour of a main crack between elastic isotropic dissimilar media under biaxial loading. A problem of considerable importance in this context is that of a crack lying along the interface of two semi-infinite elastic and isotropic bodies with different elastic properties. The plane problem of a *semi-infinite crack* between dissimilar materials was initially studied by M.L. Williams (1959) who proposed an eigenfunction approach to find the stress field near the tip of the crack. He pointed out that the stress field coming out of the elastic solution is modulated by an 'oscillatory function' of the distance from the tip. Similar problems were studied using the complex potentials technique by F. Erdogan (1963, 1965), who confirmed the oscillatory behaviour of the stress field and extended Irwin's concept of the stress intensity factor to this kind of problem. The elastic solution of the more general problem of an interfacial finite crack between two dissimilar media, loaded to infinity with a uniform stress field, was given by J.R. Rice and G.C. Sih (1965). In general, the elasticity solutions worked out by many authors show the well-known unsatisfactory behaviour at the

48

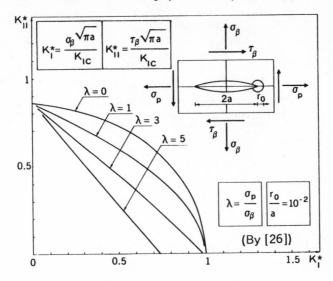

Figure 2.20   Fracture loci in non-dimensional stress intensity factor plane for some values of the parameter λ.

crack tip, i.e. the oscillatory singularities of stress and displacement components. Comnimou (1977–78) has recently shown that the oscillatory singularities can be removed by assuming that the crack is not completely open and its surfaces are in contact near the tip.

The closed-form solution of the problem in which two bonded dissimilar semi-infinite isotropic elastic media $M_j(j = 1, 2)$ containing a crack of length $2a$ along the interface and subjected at infinity to normal and shear loading, can be obtained by using the complex variable approach. By using Kolosov-Muskhelishvili equations, the boundary conditions at infinity for the M_1 and M_2 media, the traction free conditions on the crack surfaces, and traction and displacement conditions along the bond line, the corresponding potential functions can be obtained (see, e.g. [27, 28]). Then the stresses and displacements at any point in the two media can easily be obtained. This approach leads to some physical inconsistencies, as mentioned above, which are related to the oscillatory character of stress and displacement field near the tips of the interfacial crack.

However, when the two bonded media are assumed to be *incompressible* and under *plane strain* conditions, the oscillatory behaviour of the elastic fields disappears and their expressions, confining one's attention to the crack region, are considerably simplified (see form. (2.6) (2.7) in [28]).

On the basis of the above-mentioned conditions, for the polar stress components which are of relevance in this context, one has (the symbols are explained in Figure 2.21):

$$\frac{\sigma_{\vartheta\vartheta}^{(1)}}{T} \cong (k - 1) \sin^2 \vartheta + f_I(r, \vartheta) - \frac{3S}{T} f_{II}(r, \vartheta)$$

$$\frac{\sigma_{\vartheta\vartheta}^{(2)}}{T} \cong \frac{\mu_2}{\mu_1} (k - 1) \sin^2 \vartheta + f_I(r, \vartheta) - \frac{3S}{T} f_{II}(r, \vartheta)$$

(2.12)

49

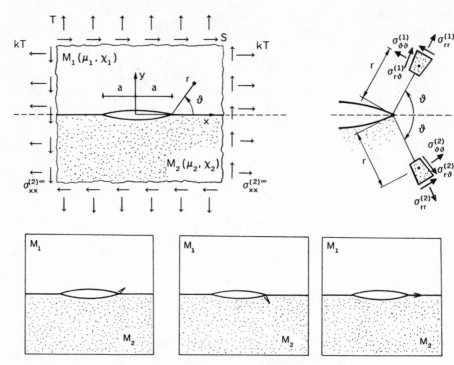

Figure 2.21a,b   Interfacial crack. Notation (a) and branching (b).

$$\frac{\sigma_{r\vartheta}^{(1)}}{T} = \frac{\sigma_{r\vartheta}^{(2)}}{T} \cong \frac{1}{4\sqrt{2}} \sqrt{\left(\frac{a}{r}\right)} \left(\sin\frac{5}{2}\vartheta - \sin\frac{\vartheta}{2}\right) +$$

$$+ \frac{1}{4\sqrt{2}} \frac{S}{T} \sqrt{\left(\frac{a}{r}\right)} \left(\cos\frac{5}{2}\vartheta + 3\cos\frac{\vartheta}{2}\right)$$

where:

$$f_I(r, \vartheta) = \frac{1}{4\sqrt{2}} \sqrt{\left(\frac{a}{r}\right)} \left(3\cos\frac{\vartheta}{2} + \cos\frac{3}{2}\vartheta\right);$$

$$f_{II}(r, \vartheta) = \frac{1}{4\sqrt{2}} \sqrt{\left(\frac{a}{r}\right)} \left(\sin\frac{\vartheta}{2} + \sin\frac{3}{2}\vartheta\right).$$

Considering that μ_1 and μ_2 are the tangential elastic moduli of the two materials, it may be seen from equation (2.12) that, as in the homogeneous case, the load 'parallel' to the crack affects only the circumferential stresses through the non-singular terms. The omission of non-singular terms could lead to error in the prediction of some quantities such as, for example, the critical applied load and the angle of incipient crack extension.

For brittle materials the Maximum Stress Criterion proposed by Erdogen-Sih [21] provides an example of a local fracture criterion allowing either the critical load levels or the branching angle to be determined.

A modified version of the Maximum Stress Criterion has been proposed in [27] to study the fracture of an elastic system consisting of two bonded dissimilar materials, under biaxial loading conditions only, with a crack along the interface. The following assumptions are made:

(1) Crack propagation will take place along the interface or into one of two adjacent materials along a direction $\vartheta_0(j)$ ($j = 1, 2$) for which the circumferential* stress $\sigma_{\vartheta\vartheta}$, evaluated at a small distance r_0 from the crack tip, attains a local maximum.

(2) Crack propagation will begin as soon as one of the following conditions is satisfied:

$$\sqrt{(2\pi r_0)}\, \sigma_{\vartheta\vartheta}^{(1)}(\vartheta_0^{(1)}, r_0) = K_{IC}^{(1)} \qquad 0 < \vartheta_0^{(1)} \leqslant \pi$$

$$\sqrt{(2\pi r_0)}\, \sigma_{\vartheta\vartheta}^{(2)}(\vartheta_0^{(2)}, r_0) = K_{IC}^{(2)} \qquad -\pi \leqslant \vartheta_0^{(2)} < 0 \qquad (2.13)$$

$$\sqrt{(2\pi r_0)}\, [\sigma_{\vartheta\vartheta}^{(2)}(0, r_0) + \sigma_{r\vartheta}^{(2)}(0, r_0)]^{1/2} = K_b \qquad \sigma_{\vartheta\vartheta} > 0$$

where $K_{IC}^{(1)}$, $K_{IC}^{(2)}$ are the critical stress intensity factors of the two materials and K_b is a critical parameter taking into account the adhesive of the bond together with the coplanar crack propagation; the debonding condition (2.13) is valid in the absence of frictional effects, i.e. when $\sigma_{\vartheta\vartheta}(0, r_0) > 0$.

The distance r_0 appearing in (2.13) should be considered to be at least the maximum length of the characteristic 'process zone'. In any case, r_0 should be considered a material property.

The above-mentioned fracture criterion states that crack extension occurs in one of the bonded media or along the bond line, depending on the fracture properties of the bonded materials, the kind of bond and the elastic properties of the bonded material (Fig. 2.21b).

In Figure 2.22 the critical values of the applied tension $T^* = \sqrt{(2\pi r_0)}\, T/K_b$ for $\mu_1/\mu_2 = 3$ and for various stress ratios S/T are represented respectively. From Figure 2.22 it is possible to know whether the crack extension occurs along the interface or into one of the materials.

Consider now an elastic matrix containing a partially bonded rigid elliptic inclusion. By considering the two symmetric partial debonds as interface cracks, a solution of the problem for biaxial load to infinity may be looked for by conformal mapping technique of the theory of complex variable functions. Figures 2.23 to 2.26 show the stress components in the elliptical coordinate system (ξ, η). The variation of the dimensionless normal stress $\bar{\sigma}_{\eta\eta} = \sigma_{\eta\eta}/N$ along the crack surface is shown in Figure 2.23, when $r = b/a = 0.2$ and for various values of $S = T/N$ as well as for the crack angle $\vartheta_1 = 20°$ respectively [29]. It may be noted that the above stress component is compressive in a small region near the middle of the crack and then is tensile up to the crack tip where it tends to infinity. Figure 2.24 shows the results considering the angle of crack ϑ variable. Figures 2.25 and 2.26 show the tensile and compressive lateral load effect on the dimensionless stresses $\bar{\sigma}_{\xi\eta} = \sigma_{\xi\eta}/N$ and $\bar{\sigma}_{\eta\eta} = \bar{\sigma}_{\xi\xi} = \sigma_{\xi\xi}/N$ respectively, along the bonded part of the inclusion ($\vartheta < \eta \leqslant \frac{\pi}{2}$), with the crack angle $\vartheta = 10°$ and $r = b/a = 0.2$.

---

*The symbol for circumferential normal stress is now denoted by $\sigma_{\vartheta\vartheta}$ instead of σ_ϑ.

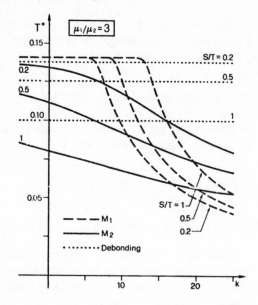

Figure 2.22   Critical values of the applied tension vs biaxial load factor K for various stress ratios and $\mu_1/\mu_2 = 3$.

Limiting attention to the crack tip and considering a local coordinate system (Figure 2.27), a definition of the stress intensity factors K_1 and K_2 can be given, as shown in [29]. Then the expressions of the stress components in an elliptic coordinate system at the vicinity of the crack tip are possible and their expressions in a local Cartesian coordinate system can be found. In this way, the fracture response of the

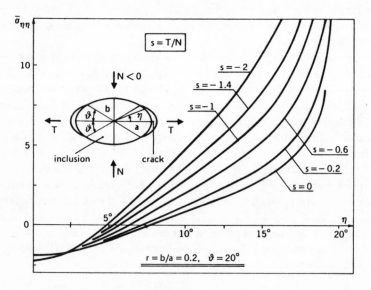

Figure 2.23   Lateral load effect of the dimensionless normal stress $\bar{\sigma}_{\eta\eta}$ along the surface of the crack.

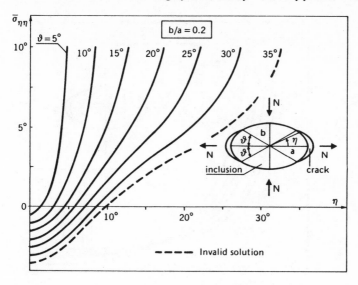

Figure 2.24   Variation of the dimensionless normal stress $\bar{\sigma}_{\eta\eta}$ along the surface of the crack for various values of the crack angle ϑ.

elastic system, i.e. the crack extension at the interface and its deviation into the matrix, may be investigated.

MODEL 1: mono-phase, two-dimensional (elastic matrix + cracks). An elementary two-dimensional cell is considered [24], in which an infinite number of micro-cracks with constant length $2a$ are embedded in an elastic matrix. The micro-cracks are randomly oriented but preserve the discontinuity surface orthogonal to the cell

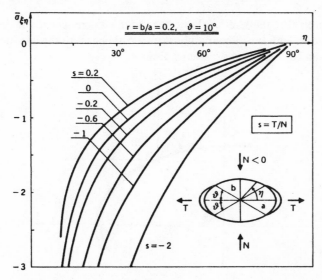

Figure 2.25   Lateral load effect on the dimensionless stress $\bar{\sigma}_{\xi\eta}$ along the bonded part of the inclusion.

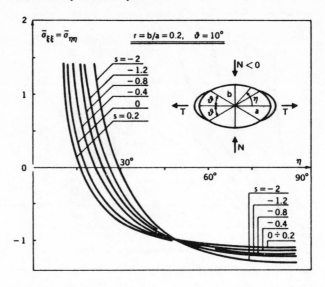

Figure 2.26   Lateral load effect on the dimensionless stresses $\bar{\sigma}_{\xi\xi} = \vec{\sigma}_{\eta\eta}$ along the bonded part of the inclusion.

middle plane (Fig. 2.28). The stress interaction between adjacent cracks is assumed non-existent in this model.

If σ_0 denotes the elastic limit of the actual material to be modelled in a mono-axial tensile test and K_{IC} the critical stress intensity factor (mode I) of the matrix, the following relationship can be written applying expression (2.2):

$$a_0 = \frac{1}{\pi} \left(\frac{K_{IC}}{\sigma_0} \right)^2 \tag{2.14}$$

on the hypothesis that the limit of elastic behaviour is a consequence of the extension of a 'critical crack' with the discontinuity surface orthogonal to the direction of stress; the length a_0 can be considered a 'material property'.

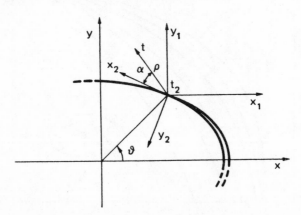

Figure 2.27   The local coordinate system.

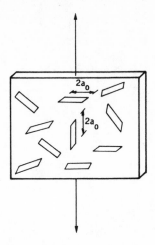

Figure 2.28   Material model.

Consider now the above-mentioned cell in a bi-axial state of stress. In the Mohr plane $\sigma_\beta - \tau_\beta$ (notation in Figure 2.29a) with the position $a = a_0$, the limit condition of crack extension can be derived from the $K_I - K_{II}$ plane limit condition [corresponding to conditions (2.3), (2.5)]. In this case, in fact, for each crack, according to relations (2.2), one has:

$$K_I = \sigma_\beta \sqrt{\pi a_0}; \qquad K_{II} = \tau_\beta \sqrt{\pi a_0};$$

then:

$$\sigma_\beta = \frac{K_I}{\sqrt{\pi a_0}}; \qquad \tau_\beta = \frac{K_{II}}{\sqrt{\pi a_0}}$$

which show that the connection between $\sigma_\beta - \tau_\beta$ and $K_I - K_{II}$ is only a matter of 'coefficient of proportionality' (ϑ_0 is constant for the model).

In the plane situation, when the principal stresses σ_1, σ_2 are known, the Mohr's circle representation (Fig. 2.29) allows one to compare the stress state to the limit condition; then the curves represented in Figure 2.29 are effectively 'Mohr's envelopes' corresponding to the elastic limit of the idealized material. Figure 2.29b shows the situation in which no crack ($2a_0$ is its length for the model), whatever its angle of inclination β, is subject to a composed stress state σ_β, τ_β, which produces its extension; the friction is also qualitatively considered.

In order to estimate propagation the mechanical behaviour of the material is defined by the critical parameters $K_{IC}/\sqrt{a_0}$ and f, where the latter is the internal friction coefficient. In this case, the traditional parameters for intrinsic curve are evaluated by 'fracture mechanics parameters'.

In order to build up an elastic domain in the geometrical plane of the two extreme principal stresses, three hypotheses can be made:

(i) the Mohr's envelope-curves, whose validity has been stated for bi-axial state of stress, are valid also for the tri-axial state of stress;

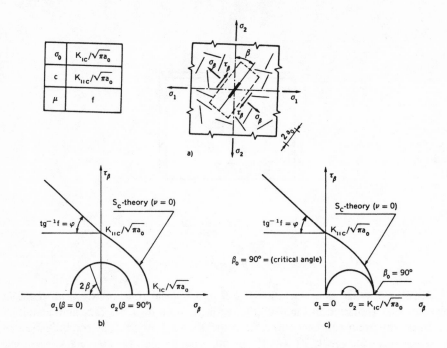

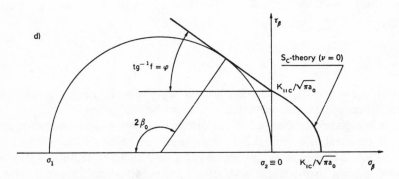

Figure 2.29   Limit curves in Mohr plane for the mono-phase two-dimensional model.

(ii) the influence of the intermediate principal stress is considered absent;

(iii) the linear relation $\sigma - \tau$ in the compression quadrant of the Mohr plane is considered valid up to the tangent point with the mono-axial critical circle in compression.

Some comments should be made about the first hypothesis. Consider a solid under general loading and extrapolate an element in the form of a parallelepiped (Fig. 2.31 a,b) wth a constant tensional state on each face and the direction of edges coincident with the principal of stresses (the faces are submitted to normal stress only). The principal stresses are σ_1, σ_2, σ_3 (the intermediate stress is σ_2), the same supposed in the corresponding element of the solid under consideration.

56

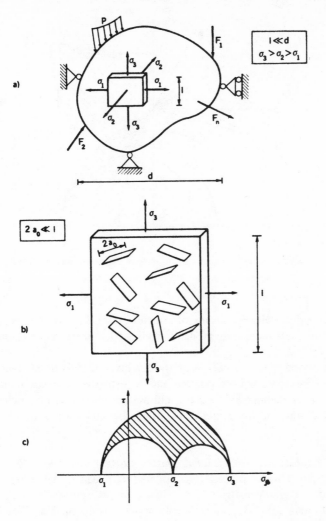

Figure 2.30   Element of the material model in the form of a parallelepiped.

If one considers a great number of micro-cracks in the element with the surface randomly oriented and length $2a_0$ smaller than the dimension of the element, the most dangerous cracks will be those with the surface orthogonal to the plane of the stresses σ_1, σ_3. In fact, these cracks are in an evidently more severe stress condition; they behave according to mode I + mode II conditions. On the basis of these considerations, the tri-axial state of stress produces an effect similar to the bi-axial σ_1, σ_3 state of stress, In Figure 2.31, four loci on the geometrical plane of extreme principal stresses (σ_1 vs σ_3) are shown for four different values of the friction angle between the crack surfaces. The concavity of the Mohr's locus (see Figure 2.18) is translated into the concavity on the locus on σ_1 vs σ_3 geometrical plane (tension-compression quadrants). Such a concavity is consistent with Drucker's Postulate, since 'friction resistance' in the material is here considered. Several authors have obtained this

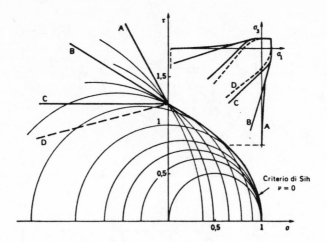

Figure 2.31   Fracture loci on the geometrical plane of extreme principal stresses for the material model.

concavity experimentally by means of concrete tests. In this case, applying the Sih criterion to the present concrete model, including friction effects, the concavity can be theoretically explained.

A similar model has also been recently developed (1979) by M. Lino [30]; using friction the concavity has been pointed out. A further mono-phase model was developed in 1979 by Bamberger, Cannard and Marigo [31] in order to characterize the micro-cracked state of the material by using the propagation speed of ultrasonic waves.

MODEL 2: Bi-phase, two-dimensional (elastic matrix + inclusions + bond cracks). An elementary two-dimensional cell is considered in which a rigid elliptic inclusion is embedded in the elastic matrix. The inclusion representing the aggregate particles is assumed infinitely rigid [32] and partially bonded to the matrix. The assumption of the rigidity could seem rather restrictive, but certainly the model provides a good approximation for studying the behaviour of normal concrete, whose aggregates are sufficiently harder with respect to the matrix.

The elementary cell is represented in Figure 2.32, where the semi-axes of inclusion are $a = c(1 + k)$; $b = c(1 - k)$ and the dimension d of the cell, in comparison with a or b, is indicative of the volume fraction of coarse aggregate. The inclusion form is then characterized by the parameters c, k. If $k = 0$, a circular rigid inclusion of radius c is derived as a particular case.

The solution of the elastic problem for biaxial load to infinity can be worked out, as previously discussed, by using the conformal mapping technique of the theory of complex variable functions, The stress components in the immediate vicinity of the crack tip can be found as functions of stress intensity factors. Moreover, the fracture response of the elastic system can be studied by using the tensional criterion.

By a simple procedure it is possible to obtain the approximate average strain components with relation to both loads and the debonded arc ϑ. In fact, the average

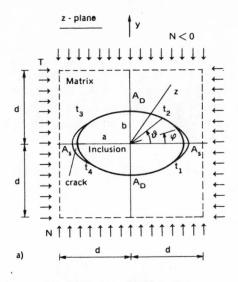

Figure 2.32   The elliptic inclusion.

strain energy per unit volume corresponding to the model of Figure 2.32 can be written as follows:*

$$W = W_0 + W_c(\vartheta)$$

where W_0 is the average strain energy for the cell with no cracks and $W_c(\vartheta)$ is the strain energy contribution due to cracks. Under the cell in monoaxial compression, the average strain can be derived by using the Castigliano theorem:

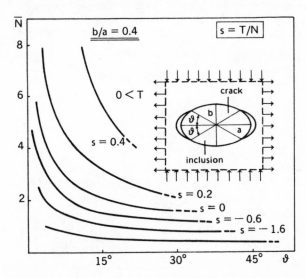

Figure 2.33   Lateral load effect on the critical compressive load N corresponding to 'debonding'.

*Note the different meaning of ϑ compared to the previous part of this chapter.

$$\epsilon(\vartheta) = \frac{\partial W}{\partial N} \tag{2.15}$$

The detailed expression of equation (2.15) has been worked out in [32]. The stress strain (2.15) is linear, for a fixed value ϑ_c of the crack angle ϑ, and will terminate at a critical value $N = N_c$ which causes the debonding. It follows that a non-linear response, due to progressive bond separation, can be obtained by considering the sequence of critical points N_c, $\epsilon_c = \epsilon(\vartheta_c)$ on the linear response curves corresponding to successively larger values $\vartheta = \vartheta_c$.

On the basis of the debonding criterion previously described, it is possible to evaluate the critical compressive load N_c. If one writes the dimensionless critical compressive load in the form:

$$\bar{N} = N_c \frac{\sqrt{\pi c}}{K_b} \tag{2.16}$$

K_b being the critical value of the interfacial stress intensity factor, it will be possible to plot this against the crack angle, for various values of the applied stress ratio $s = T/N$. The relationship evaluated [32] in the particular case $b/a = 0.4$ is shown in Figure 2.33.

The effect of the lateral tensile stress and the effect of the aggregate size on the stress-strain response is shown in Figure 2.34. The linear response curves with no cracks ($\vartheta = 0$) are valid up to those values of N for which crack formation occurs. If the bond strength is specified as σ_b, at a compressive load N_c such as:

$$\sigma_{\xi\xi}(\vartheta = 0, \varphi = 0) = \sigma_b$$

bond failure is expected to start at $\varphi = 0$.

The debonding phenomenon here analyzed is the one responsible for non-linearity

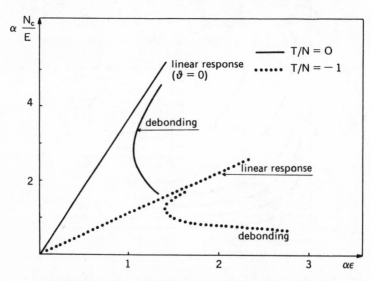

Figure 2.34   Lateral tensile stress effect on the stress-strain response for $c = 1$ and $b/a = 1$.

in the stress-strain relation corresponding to the interval of stress III (Fig. 2.3). The actual concrete could be seen as a set of elementary cells like that of this model, in which the crack angle ϑ at the interface in randomly different as far as location is concerned.

A homogeneous matrix with a polygonal inclusion representing an aggregate particle is now considered [33]. An initial interfacial crack with length $2l_1$ is assumed to be located along one side (AB, see Figure 2.35) of the inclusion. This crack spreads (Mode II) in an unstable way as soon as a critical load is reached. This shear crack reaches the length $2L_1$ (equal to the aggregate side length) and then stops (Fig. 2.35b), because further crack propagation in the same inclined direction would take place through the matrix. If the external load is increased to a higher value, branching cracks in the matrix will develop (Fig. 2.35c). This approach [33] provides the actual crack length in the matrix as a function of loading.

The possibility that the branching crack AA' meets a second inclusion as it propagates is considered in Figure 2.36. Further crack growth will take place either through the inclusion, thus maintaining the same direction, or follow the interface MN.

Three different cases can be considered: normal concrete, high strength concrete and concrete with lightweight aggregates.

On the basis of the previous statements, it is possible to simulate concrete damage using the Monte-Carlo Method. A typical example of computer simulation of concrete is shown in Figure 2.37 with 30 polygonal inclusions. Each particle is supposed to have one interfacial crack. As load increases, the most critical cracks will propagate at first. A further increase of the load produces a characteristic crack pattern (Fig. 2.38) until one crack finally runs through the total specimen. This phenomenon is defined as 'failure' of the concrete. Then the gradual damage of concrete and its failure are explained in this model by means of Fracture Mechanics concepts.

Concrete models for damage evaluation under time-depending loads. Damage accumulation in time in concrete structures under cyclic external load is generally controlled by laws which are not completely known or which cannot be formulated in a deterministic manner. Thus damage can be described only referring to *stochastic process concepts.*

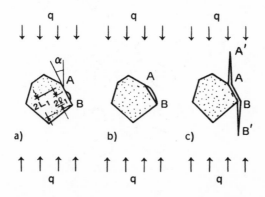

Figure 2.35   Initial debonding crack grows in the matrix [33].

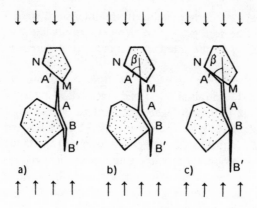

Figure 2.36   Typical crack path [33].

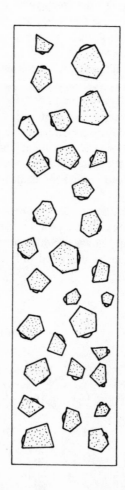

Figure 2.37   Typical computer realization of the random structure of normal concrete [33].

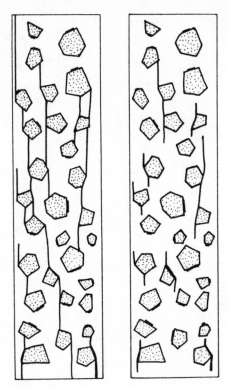

Figure 2.38   Crack pattern for two different load levels in normal concrete [33].

The state of deterioration (damage) of an elementory volume of concrete in a structural element under time-depending loading can be described as a 'system which evolves through a finite number of intermediate states corresponding to the whole numbers 2 . . . $(r-1)$ and comprised between the initial state 1 and the final state r'. In the initial state the system is assumed to be able to perform certain functions in a specified time interval. In this regard, the behaviour of a family of random variables $\{x_t,\ t{\in}T\}$ should be known specifying the value of the time-depending damage parameter.

In this context, the gradual deterioration of concrete caused by micro-cracking under time-dependent loading conditions has been characterized by considering this process as a Markoff-chain process. Four states of damage [37] are represented in Figure 2.39. Some applications of the Markoff-chain process to materials have been developed over the last few years [34–36] which have shown promising results.

A different approach to evaluate theoretically the progressive deterioration of materials under monotonic or cyclic loads is the 'damage theory' introduced by Kachanov (1959). This theory was developed more in detail by Rabotnov (1968) and Broberg (1974). An approach combining 'damage theory' and 'Fracture Mechanics' was proposed in 1977 by Janson and Hult [38].

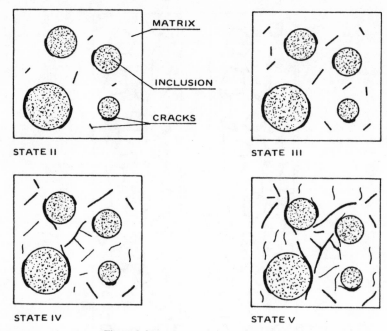

STATE II   STATE III

STATE IV   STATE V

Figure 2.39   Four states of damage.

References

[1]   Robinson, S.R., Methods of Detecting the Formation and Propagation of Microcracks in Concrete, *Proceed. of the Int. Conf. 'The structure of concrete'. London, 1965.*

[2]   Slate, F.O. and Olsefski, S., X-rays for Study of Internal Structure and Microcracking of Concrete, *Journ. of American Concrete Institute, Proceed.,* 60, pp. 575–578 (1974).

[3]   Dhir, R.H. and Sangha, M., Development and Propagation of Microcracks in Plain Concrete. *Matériaux et Constructions,* 37, pp. 17–23 (1974).

[4]   Mamillan, M. and Bouneau, A., Nouvelles applications des mesures de vitesse du son aux matériaux de construction, *Annales de l'ITBTP,* série EM-178 (April 1980).

[5]   Terrien, A. and Bergues, J., Study of Concrete's Cracking under Multiaxial Stresses. 5th *Int. Conf. on Fracture,* Cannes, (1981).

[6]   Mazars, J., Evolution of Microcracks in Concrete: the Formation of Cracks, *Annales de l'ITBTP,* série Béton – 202, October 1981.

[7]   Tognon, G.P., Ursella, P. and Coppetti, G., Bond Strength in Very High Strength Concrete, *Int. Congr. on the Chemistry of Cement,* Paris, (1980).

[8]   Shah, S.P., Chandra, S., Critical Stresses, Volume Change and Microcracking of Concrete, ACI *Journal,* (Sept. 1968).

[9]   Beres, L., Relationship of Deformational Processes and Structure Change in Concrete, *Struct., Solid Mech and Eng. Design,* Part 2 (Ed. TE'ENI), Wiley-Interscience, (1971).

[10] Newman, K., Newman, J.B., Failure Theories and Design Criteria for Plain Concrete, *Struct., Solid Mech. and Eng. Design,* Part 2 (Ed. TE'ENI), Wiley-Interscience (1971).

[11] Di Leo, A., Di Tommaso, A., Merlari, R., Danneggiamento per microfessurazione di malte di cemento e calcestruzzi sottoposti a carichi ripetuti. *La Prefabbricazione* (Italian Journ.), 11, pp. 577–587 (1979).

[12] Santiago, S.D., Hilsdorf, H.K., Fracture Mechanics of Concrete under Compressive Loads, *Cem. Concr. Research,* 3, pp. 363–388 (1973).

[13] Mehmel, A., Kern, E., Elastiche und Plastiche Stauchungen von Beton infolge Druckschwell und Standbelastung, *Deutsch Ausschuss für Stahlbeton*, 153, Berlin (1962).

[14] Cervenka, V., Behaviour of Concrete under Low Cycle Repeated Loadings, AICAP-CEB *Symposium*, vol. 2, Rome (1979).

[15] Evans, R.H., Marathè, M.S., Microcracking and Stress-Strain Curves for Concrete in Tension, *Matériaux et Constructions*, 1 (1968).

[16] Heilman, H.G., Hilsdorf, H.H., Finsterwalder, K., Festigkeit und Verformung von Beton unter Zugspannungen, *Deutsch. Auschuss für Stahlbeton*, 203, Berlin (1969).

[17] Hillerborg, A., A Model for Fracture Analysis, *Report* TVBM 3005, Lund (Sweden), 1978.

[18] Terrien, M., Emission acoustique et comportemen post critique d'un béton sollicité en traction, *Bulletin de Liaison des Ponts et Chaussées*, No 105, pp. 65–72 (1980).

[19] Di Leo, A., La prova di trazione diretta del calcestruzzo, *Ingegneri, Architetti, Costruttori (INARCOS)*, Bologna, 408 (1980).

[20] Kupfer, H., Hilsdorf, H.K., Rüsch, H., Behaviour of Concrete under Biaxial Stresses, A.C.I. *Journal*, Proceed., 66 (August 1969).

[21] Erdogen, F., Sih, G.C., On the Crack Extension in Plates under Plane Loading and Transverse Shear, *Journ. of Basic Engineering* (Dec. 1963).

[22] Sih, G.C., Some Basic Problems in Fracture Mechanics and New Concepts, *Eng. Fracture Mechanics*, 5, 1973.

[23] Di Tommaso, A., Nobile, L., Viola, E., Diramazione di un crack dominante in un solido a regime deformativo biassiale, III *Congr. Na.* A.I.M.E.T.A. (*Ass. Italiana di Meccanica Teorica e Applicata*), Cagliari (Oct. 1976).

[24] Carpinteri, A., Di Tommaso, A., Viola, E., Stato limite di frattura nei materiali fragili, (Modelli Meccanici Teorici), *Giornale del Genio Civile* (Italian), fasc. 4–5–6, pp. 201–224 (1978).

[25] Eftis, J., Subramonian, N., Liebowitz, H., Crack Border Stress and Displacement Equations Revisited, *Eng. Fracture Mechanics*, 9, pp. 189–210 (1977).

[26] Carpinteri, A., Di Tommaso, A., Viola, E., Collinear Stress Effect on the Crack Branching Phenomenon, *Matériaux et Constructions*, 12, pp. 439–446 (1979).

[27] Viola, E., Piva, A., Biaxial Load Effects on a Crack between Dissimilar Media, *Eng. Fracture Mechanics*, 13 pp. 143–1 (1980).

[28] Viola, E., Piva, A., Plane Strain Interfacial Fracture Analysis of a Bimaterial Incompressible Body, *Eng. Fracture Mechanics*, 15, (1981).

[29] Viola, E., Piva, A., Fracture Behaviour by two Cracsk around an Elliptical Rigid Inclusion, *Eng. Fracture Mechanics*, 15, pp. 303–325 (1981).

[30] Lino, M., Modèle de béton microfissuré, *Seminaire*, 'Modèle de comportement de béton fissuré', *Ecole Polytechnique, Palaiseau* (1979).

[31] Bamberger, Y., Cannard, G., Marigo, J., Microfissuration du béton et propagation d'ondes ultrasonnes, *Euromech.* 115 (Anisotropic), Grenoble (1979).

[32] Piva, A., Viola, E., Stress Strain-Response of a Concrete Mathematical Model, *Proc.* AIMETA (*Ass. Italiana di Meccanica Teorica e Applicata*), Palermo, Italy (1980).

[33] Zaitev, Y.B., Wittmann, F.H., Simulation of Crack Propagation and Failure of Concrete, *Materials and Structures* (RILEM Journ.), 83, pp. 357–365 (1981).

[34] Chow, T.H., Shah, H.S., Markov Process Model for Creep of Concrete under Constant Sustained Compressive Stresses, *Struct., Solid Mech. and Eng. Design*, (Ed. TE'ENI), vol. 1, Wiley-Interscience, London (1971).

[35] Mihashi, H., Izumi, M., A stochastic Theory for Concrete Fracture, *Cem. Concr. Research*, 7, (1977).

[36] Mihashi, H., Wittmann, F.H., Stochastic Approach to Study the Influence of Rate of Loading on Strength of Concrete, *Heron Publications* (The Netherlands), Vol. 25, No 3 (1980).

[37] Viola, E., Un approccio stocastico allo studio del danneggiamento del calcestruzzo prodotto dall'azione dei carichi ripetuti, *Giornale del Genio Civile* (Italian), 1983 (in press).

[38] Janson, J., Hult, J., Fracture Mechanics and Damage Mechanics A Combined Approach, *Journ. de Mécanique Appliquée*, 1, pp. 69–84 (1977).

65

S. MINDESS

3

Fracture toughness testing of cement and concrete

3.1 Introduction

While the cracking of hardened cement paste and concrete has been studied extensively at least since the work of Richart et al [1] in 1928, it was only about thirty years later that the concepts of fracture mechanics (first proposed by Griffith [2] in 1920) were finally applied to cementitious materials. However, in spite of a great deal of work since 1959, resulting in over four hundred published research reports on the fracture of cement and concrete [3], there is still much controversy as to the very applicability of fracture mechanics principles to these materials. Considering the many uncertainties that yet exist about the precise nature of the cracking process, and indeed even about the notch sensitivity of these materials, it should not be surprising that there is still no agreement as to which fracture mechanics criteria best characterize the failure of cement or concrete, or which test methods are most suitable. The purpose of the present work is to review the applications of fracture mechanics to hardened cement paste (hcp), concrete, fibre reinforced concrete (frc), and polymer impregnated concrete concrete (PIC). The focus will be on the fracture toughness parameters and techniques that have been used till now, and on the difficulties that have arisen in trying to define valid specimen sizes and test geometries.

3.2 Physical phenomena involved in fracture

In order to develop realistic fracture models for concrete, a knowledge of the physical processes involved in crack propagation is necessary. This requires an understanding of the material behaviour at two different levels: on the microstructural scale, the mechanism of crack growth through hydrated cement paste; on the macroscopic scale, the development of cracks in a system in which aggregate particles, and possibly also fibres, are dispersed in a continuous matrix of hydrated cement paste;

Cracks in Hardened Cement Paste. The mechanism of crack growth in hardened cement paste is still not very well understood. At early ages, up to perhaps several weeks, it appears that cracks propagate preferentially through the high porosity

C–S–H (calcium silicate hydrate) phase [4,5], since at early stages of hydration both the unhydrated cement particles and the calcium hydroxide that is one of the early reaction products correspond to low porosity regions and act as rigid inclusions. This is shown in Figure 3.1 [6,7]. The unhydrated cement grains and the $Ca(OH)_2$ crystals often even seem to act as crack arrestors. However, in mature pastes, this discrimination is lost as the matrix becomes less porous and more homogeneous, and the fracture path becomes more direct. There is also some evidence [8,9] that even in mature pastes, fracture occurs preferentially along the weakly bonded basal planes of the $Ca(OH)_2$. This may be due to the fact that some preliminary data indicate [10] that the tensile strength of $Ca(OH)_2$ is only about 1 MPa, which is considerably less than that of ordinary cement paste.

The cracking of hardened cement paste has been studied in detail using a scanning electron microscope (SEM) [11–13]. Both a wedge-loaded compact tension device [11] and a compression device [12,13] have been developed which allow a specimen to be tested directly in the sample chamber of the SEM, so that cracking can be monitored under load. Using the compact tension specimen [11], it has been noted that while the cracks tend to be fairly straight, they actually seem to consist of a linked series of short segments that zig-zag back and forth around the direction of propagation. The cracks are approximately, but not completely, parallel sided. At least on the surface that can be observed, they are also sometimes not continuous. For instance, in Figure 3.2a, the trace of the crack appears to terminate, and a seemingly parallel crack displaced from it occurs and continues onward. Some crack branching is also observed (Figure 3.2b), particularly as the apparent crack tip is approached. Where crack branching does occur, only one branch will remain 'active' under further loading.

In compression [12], much the same features are observed. Cracks tend to run approximately parallel to the axis of loading, and a certain amount of crack branching occurs. As shown in Figure 3.3a, the cracks generally appear to go around unhydrated cement grains, which act as microaggregates. However, cracks occasionally do go through such particles, as shown in Figure 3.3b.

The process of cracking in hardened cement pastes has also been modelled by Zaitsev and Wittmann [14] using computer simulations. In this model (Figure 3.4) the cracks are assumed to originate at the capillary pores, and bridge between pores as the load increases.

Cracks in mortar and concrete. Due to the presence of the aggregate particles, the cracking processes in mortar and concrete are more complicated than those in hardened cement paste. The σ-ϵ curves for normal aggregates and for hardened cement paste are linear almost to the point of failure, but the σ-ϵ curve for concrete is quite nonlinear. This non-linearity is due largely to the very imperfect bond that exists between the cement and the aggregate, and to the progressive microcracking that takes place, primarily at the cement-aggregate interface, as the load increases. (The much more linear σ-ϵ curve for cement is probably due to the much better bond between the hydration products and the remaining unhydrated cement grains). In addition, it has been shown [15–17] that the presence of hard aggregate particles in a softer cement matrix will alter the local strain and stress distributions which lead to crack propagation and failure.

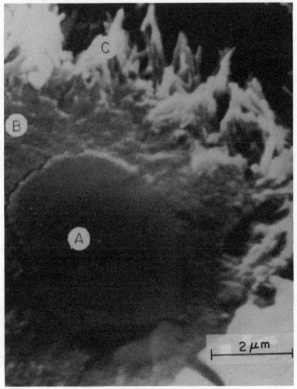

Figure 3.1   (a) Cracking in a C_3S paste hydrated for 14 days. The gray areas are $Ca(OH)_2$. The cracks ('lightening bolts') appear to be largely in the C–S–H phase. Micrograph courtesy of Dr. R.L. Berger [6]. (b) Fracture through a partially hydrated C_3S grain. A = unhydrated core; B = inner hydration production; C = acicular hydration products. Note the crack around the unhydrated core. Micrographtesty of Dr. J.F. Young [7].

69

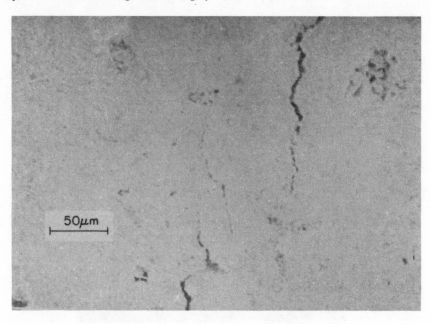

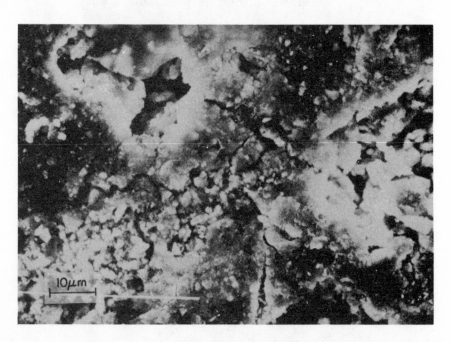

Figure 3.2.   (a) Discontinuous cracking in hardened cement paste [11]. (b) Crack branching in hardened cement paste.

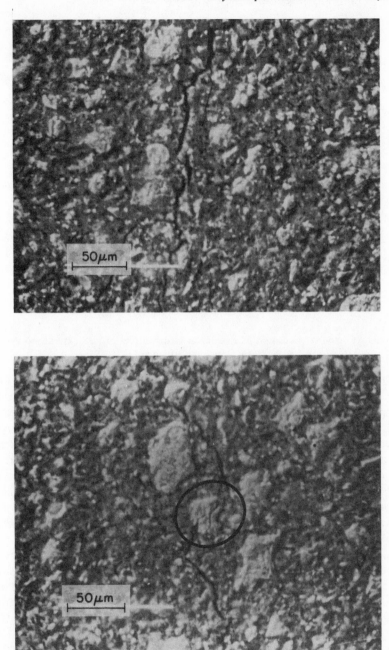

Figure 3.3   (a) Cracks generally go around 'microaggregate' particles (unhydrated cement grains);
(b) Crack going through unhydrated cement grain (circled) [12].

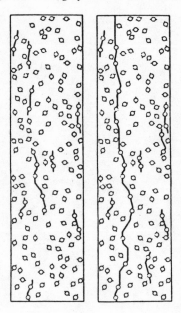

Figure 3.4.   Computer simulation of cracking in hardened cement paste; cracks are assumed to originate at the capillary pores [14].

There is general agreement that the cement-aggregate interface is the weakest region of the concrete. Even prior to loading, cracks have been observed at the cement-aggregate interface [18], due to the bleeding, volume changes during hydration, and drying shrinkage. Up to about 30% of the ultimate stress (σ_{ult}), there is very little extension of these cracks for normal rates of loading, and the σ-ϵ curve is essentially linear. Beyond $0.30\,\sigma_{ult}$, the cracks begin to grow under increasing load, partly due to the differences in the elastic constants for aggregate and for hydrated cement paste, and partly due to the high stress concentrations at the cement-aggregate interface [19], and the σ-ϵ curve becomes increasingly non-linear. After about $0.5\,\sigma_{ult}$, the bond cracks also begin to extend through the hardened cement matrix, bridging between the coarse aggregate particles, but still in a stable fashion. Finally, beyond about $0.75\,\sigma_{ult}$, the matrix cracks begin to form a much more extensive network, but there is still enough redundancy in the system for it to remain reasonably stable under short-term loading. Eventually, this crack network becomes so extensive that failure occurs, though it should be noted that even at σ_{ult} the crack pattern still leaves a structure with some load-bearing capacity; with a sufficiently stiff testing machine, the σ-ϵ curves in both tension [20] and compression [21] show a considerable descending branch. Static fatigue in concrete is associated with loading beyond about $0.75\,\sigma_{ult}$, presumably due to slow crack growth both at the interface and in the matrix. For sustained stresses below about $0.75\,\sigma_{ult}$, delayed failure does not occur in compression, probably because whatever slow crack growth does occur is compensated for by strengthening due to the consolidating effect of the load.

The fact that cracking is associated with the cement-aggregate interface has been observed using optical microscopy by Hsu et al. [18], and by Mindess and Diamond

[12, 22, 23]. Examining mortar specimens in the SEM under load, it is seen that cracks develop preferentially at the sand-cement interface, often where some ill-defined shrinkage cracks appeared to pre-exist. The crack surface is very tortuous, and a considerable amount of branch and multiple cracking is noted. It appears that the cracking process is more complicated than is assumed in most of the simple fracture mechanics models which have been applied to mortars and concretes. Some of the features of these cracks are shown in Figure 3.5.

Crack growth in concrete has also been modelled by Zaitsev and Wittmann, as described earlier for cement [14]. In this model, cracks are assumed to originate at the cement-aggregate interface. As shown in Figure 3.6, cracks in ordinary concrete are modelled as going around aggregate particles. For lightweight concretes, where the aggregate particles are weaker than the cement matrix, the cracks go right through the aggregate particles; for high strength concrete, which is often characterized by a very good cement-aggregate bond, some cracks are also assumed to pass through the aggregate particles.

Cracks in Cement Composites. When discrete, randomly oriented fibres are introduced into concrete, or when concrete is impregnated with polymers, the situation becomes even more complex. Not only is there an additional phase introduced into the system, but there is also a fundamental change in the material behaviour. Fibre additions improve the 'toughness' (or area under the σ-ϵ curve to failure); they give concrete a considerable amount of apparent ductility. Polymer impregnation generally improves both the elastic modulus and the toughness of the concrete.

Turning first to polymer impregnated concrete (PIC), the mechanisms by which polymer impregnation improve the toughness are not completely understood. Evans et al [24] have suggested that the polymers tend to bridge across cracks, thereby increasing the resistance to crack extension. On the other hand, it has been argued [25, 26] that it is by decreasing the porosity, hence eliminating some stress-inducing flaws, that the tendency for crack growth is reduced. It has also been suggested [27] that polymer impregnation acts by improving the cement-aggregate bond. In addition, using high-speed photography, Bhargava and Rehnstrom [28] found that the crack velocity in PIC under explosive loading is about 800 m/s, compared to 180 m/s for plain concrete, indicating that crack tortuosity and microcracking are reduced in PIC.

The exact role of the fibres in fibre reinforced concrete (frc) is still not completely clear. Fibres tend to bridge across cracks, increasing the resistance to further crack growth. It has also been suggested [29] that the function of steel fibres is effectively to reduce K_I for a given stress level, due to the restraint of the adjacent fibres. Additional energy is required to strip the fibres from the matrix ahead of the crack tip [30, 31]. Aleszka and Schnittgrund [32] found that the presence of fibres always increases the irregularity of the macroscopic fracture surface, which reveals cracking at both the cement-aggregate and cement-fibre interfaces. They postulate that in frc, fracture begins with cracks at the cement-aggregate interface; when they propagate into the matrix, they are obstructed by the fibres, increasing the energy requirement for fracture. As with plain concrete, Swamy [33] observed considerable branch cracking during crack extension, while Patterson and Chan [34] noted a large zone

73

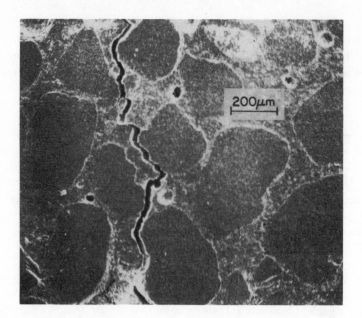

a

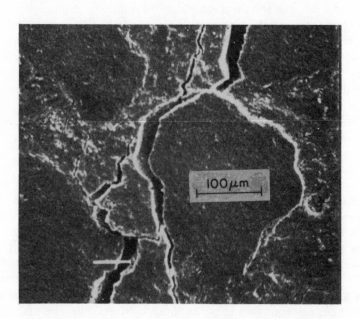

b

c

Figure 3.5.   Crack patterns of mortar loaded within an SEM: (a), (b) compact tension specimens; (c) compression specimen. Note that the cracks tend to go around the aggregate particles.

of fine cracks around the macroscopic crack tip, which seemed to move with the crack tip as the crack extended.

3.3 Rate of loading effects

The σ-ϵ behaviour and the strength of hardened cement paste and concrete are sensitive to the loading rate. As the rate of loading is increased, the σ-ϵ curve becomes more linear, and E increases [35]. In addition, typically, the compressive strength of concrete will approximately double when the strain rate is increased by six orders of magnitude [36], while the flexural strength of cement and mortar may increase by about 30% with a similar increase in strain rate [37]. However, the total strain to failure in these flexural tests was essentially constant.

The mechanisms responsible for this behaviour in cement and concrete are also not fully understood. Presumably, slower loading rates allow more subcritical crack growth to occur, thus leading to the formation of larger flaws, and hence smaller fracture loads. Therefore, these phenomena may be explained using fracture mechanics concepts. On the other hand, it may be that slower loading rates allow more creep to occur, which will increase the total strain at a given load. The observed phenomena can then be explained in terms of a limiting value of strain to cause failure; the same results (e.g. [34]) could support a maximum strain failure criterion for concrete, as well as a fracture mechanics criterion.

75

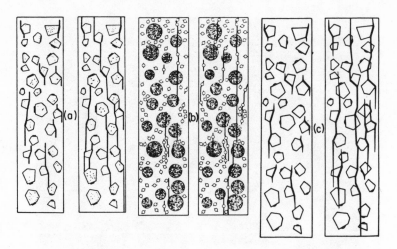

Figure 3.6   Computer simulation of cracking in concrete [14]: (a) normal concrete: cracks go around the aggregate; (b) lightweight concrete: cracks go through the aggregate; (c) high strength concrete: some cracks go through the aggregate.

3.4 Notch sensitivity

A material may be considered to be notch sensitive if the presence of a notch causes a change in the net section strength (σ_{net}) of the material (calculated on the basis of the reduced cross-section, but neglecting the stress concentrating effect of the notch). For metals, the introduction of a notch may lead to notch strengthening for highly ductile materials, due to the increased plastic constraint, or to notch weakening for materials with a limited deformation capacity, due to the stress concentrating effect of the notch. For cementitious materials, however, the notch strengthening effect is not found, and so the term notch sensitivity refers only to the possible reduction in σ_{net} due to the presence of a notch. It is then assumed that materials which are *not* notch sensitive, or in which the largest crack is less than the critical crack length, may be analyzed in terms of classical mechanics; materials which *are* notch sensitive should be analyzed using the principles of fracture mechanics. Thus, Ziegeldorf et al [38] have argued that notch sensitivity is a necessary, though not a sufficient, condition for the applicability of linear elastic fracture mechanics (LEFM). They derived the following expression for notch sensitivity (in flexure):

$$\frac{\sigma_{net}}{\sigma} = \frac{K_{IC}}{\sigma} \frac{1}{\sqrt{a}\ \left(1 - \frac{a}{b}\right)^2 F(a/b)} \leqslant 1 \tag{3.1}$$

where σ is the strength of an unnotched specimen, a is the crack length, b is the beam depth, and $F(a/b)$ is a parameter which depends upon the specimen geometry and the method of loading. This equation predicts an increase in notch sensitivity with increasing specimen size, and with an increasing ratio of σ/K_{IC}.

Unfortunately, there is still some uncertainty as to whether cementitious materials are truly notch sensitive. There are really two related questions:

1. Can the strength of hardened cement paste or concrete be controlled by artificially introduced notches, or are they always controlled by the inherent structure of these materials?

2. If artificially induced flaws can control the fracture of these materials, is there some minimum specimen size, or ratio of flaw size to specimen size, that is required before a fracture test can be considered to be valid?

The second question will be discussed at length later. Turning to the first question, a number of experimental studies of notch sensitivity have been carried out, mostly using rather small specimens. For hardened cement paste itself, there is general agreement that it is a notch sensitive material [38–45]. However, Kesler et al. [46], from tests on plates with a center crack, concluded that hardened cement paste was not notch sensitive (though a reanalysis of their data by Saouma et al [47], using modern computational techniques, suggests that their conclusions are open to question). Indeed, the notch sensitivity of hardened cement paste has been confirmed in a most interesting way by Birchall et al [48]. They determined the flexural strengths of hardened cement pastes, and related these through the Griffith equation to the sizes of the largest natural or artificially induced flaws in the material. Normal cements were found to fit the Griffith curve down to a flaw size of about 1 mm, which they found by microscopic examination to be about the size of the largest naturally occurring flaws. Cement pastes which were specifically processed to eliminate the large voids achieved much higher strengths, and fit the Griffith curve till the largest observed flaws (pores) were about 90 μm in diameter. Only below this size did other flaws in the microstructure control the strength. These results are shown in Figure 3.7.

For mortars and concretes, the results are much more contradictory. Three studies [49–51] were carried out on the effect of fairly large holes of various shapes in

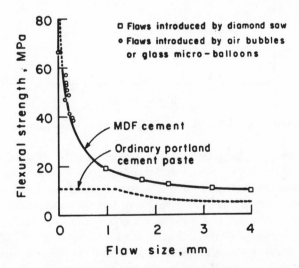

Figure 3.7   Results of flexural strength of notched MDF (macro-defect free) cement compared with ordinary portland cement paste [48].

tensile specimens. It was concluded that these holes, up to about 140 mm in diameter, had no measurable effect on concrete strength, due in part to stress redistribution near the edges of the hole. It was also argued [49] that inherent flaws in the concrete structure induced stress concentrations that could not be offset by artificially introduced flaws. A number of other investigations have also been shown that mortar and concrete are notch insensitive for notches of lengths up to at least 50 mm [39, 44, 46, 52, 53]. On the other hand, some studies have shown that mortar and concrete *are* notch sensitive, [38, 40, 54, 58], though to a lesser degree than cement.

For more complex cement composites, the results are also contradictory. Some results [44] show that both glass fibre reinforced concrete and steel fibre reinforced concrete (sfrc) are notch insensitive, while other results [59] show that sfrc is notch sensitive. Mindess and Bentur [60] have also shown that wood fibre reinforced concrete is notch insensitive in the saturated state, and at best only very slightly notch sensitive in the air-dry state. In addition, Cook and Crookham [57] have shown that polymer impregnated concrete (PIC) is notch sensitive. Some of the results described above are shown graphically in Figure 3.8.

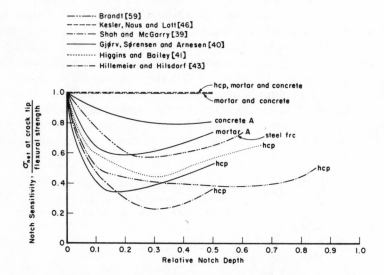

Figure 3.8.   Effect of relative notch depth on notch sensitivity.

Some of these contradictions may be explained in terms of specimen size. Walsh [61, 62] has argued that for concrete, flexural specimens would appear to be notch insensitive if they are too small, and suggested a minimum beam depth of about 230 mm. Kishitani and Maeda [63] have shown that for concrete compression specimens with an inclined crack, the material was notch insensitive for 20 mm cracks, but became notch sensitive for 40 mm cracks, with a 10−20% strength reduction. Similarly, Mai et al [64] found that asbestos-cellulose fibre reinforced cement was notch-insensitive for beam depths up to 25 mm, and that valid fracture toughness measurements could be made only when the specimen depth was greater than 50 mm. Halvorsen [65] showed that steel frc was notch insensitive for 76 mm deep specimens, but became notch sensitive for 152 mm deep specimens. In a more general way, Carpinteri [66] has also argued that specimens that are too small would be notch insensitive. While showing that for large enough specimens both mortar and concrete were notch sensitive, he defined a test brittleness number, s,

$$s = K_c/\sigma_u b^{1/2} \tag{3.2}$$

where σ_u is the ultimate tensile strength and b is the specimen depth. For $s \geqslant 0.5$, the material is no longer notch sensitive. This would imply a minimum specimen depth for notch sensitivity of about 100 mm for cement paste, and about 650 mm for concrete. (Other aspects of the problem of specimen size will be discussed later).

In summary, the notch sensitivity of cementitious materials is still an open question. Almost all investigators agree that hardened cement paste is notch sensitive. However, for concrete and fibre reinforced concrete, the results are contradictory, though the weight of the evidence suggests that even these materials are notch sensitive for specimens that are large enough.

3.5 Historical review of the applications of fracture mechanics to cementitious materials

It took almost forty years for the energy-balance concept of fracture proposed by Griffith [2] to be applied to cement and concrete. Neville [67], in 1959, suggested that the effects of specimen size on concrete strength could be related to the random distribution of Griffith flaws. The first experimental study appears to be that of Kaplan [52] in 1961, who concluded that 'the Griffith concept of a critical strain-energy release rate being a condition for rapid crack propagation and consequent fracture, is applicable to concrete'. Since Kaplan's pioneering work, a great deal of research has been carried out in this area, as has been reviewed by Mindess [68, 69].

The application of Griffith theory to concrete was developed in detail by Glucklich [70–73], who showed that the critical strain energy release rate (G_c) was much greater than twice the surface energy of concrete (2γ) because the fracture of concrete was not limited to the growth of single crack; rather, a zone of microcracking developed near the apparent crack tip, thereby creating a much larger fracture surface than the one calculated simply from the cross-sectional dimensions of the specimen. He also showed that high strength areas in the concrete, such as aggregate particles, could act so as to increase the energy demand, by forcing the crack either to go through a stronger region (requiring more energy), or to go around it (increasing the length of the crack path). Thus, due to microcracking and the influence of aggregate particles, the energy demand should increase as the crack grows.

Subsequent work (to be described in more detail below) focussed on the effects of the various concrete parameters on K_c or G_c. It was found that K_c increased with aggregate volume, with increasing size of the coarse aggregate, and with increasing roughness of the aggregate. K_c decreased with increasing w/c ratio, and with increasing air content. K_c increased with age, though, unlike stength, it seemed to reach its maximum value within about one month. K_c was also shown to increase slightly with an increase in the rate of loading. The introduction of continuous steel wires, or randomly dispersed discrete fibres of steel, glass, asbestos, and various other materials, also increased K_c, as did the introduction of polymer impregnation of the concrete.

In addition, a number of studies of subcritical crack growth in cement, concrete and fibre-reinforced concrete were carried out to determine the relationship between the crack velocity and K_I. However, there is no agreement on the best method of analyzing the data so obtained.

Another extensive series of experiments was carried out to try to determine the effects of specimen geometry and size on the fracture parameters. As will be seen below, these results too have been highly contradictory. Indeed, the variability in measurements of K_c reported in the various studies has led a number of investigators to question the applicability to LEFM to cementitious materials. For example, Kesler et al [46] concluded that 'the concepts of linear elastic fracture mechanics are not directly applicable to cement pastes, mortars and concretes'. Similar conclusions have been reached by others [50, 58, 74]. Nevertheless, most investigators continue to believe in the applicability of fracture mechanics to these materials.

Partly in response to the difficulties in applying LEFM to concrete materials, a number of non-linear fracture criteria have also been suggested: the J-integral, R-curve analysis, and the concept of a critical COD. In addition, a 'smeared' crack model has been suggested, in which the crack front is assumed to consist of a diffuse zone of microcracks, the size of which is related to the maximum aggregate size. Finally, the 'fictitious' crack model has also been developed; this is a tied-crack model in which stresses are assumed to act across a crack as long as it is only narrowly opened. However, in spite of all of this work, there is still no agreement as to the 'best' way of characterizing the fracture of cementitious materials.

3.6 Fracture toughness parameters and techniques

Over the years, a great many different techniques have been used to determine the fracture properties of cementitious materials. This reflects the fact that, as yet, there are no tests that have been standardized for cement and concrete. Consequently, it is very difficult to compare the results obtained by different investigators, since they would typically have been obtained on specimens of different types and sizes, and on very different sorts of testing machines; curing conditions have also not been standardized. This may account for much of the scatter in the values reported below. It is perhaps most convenient to summarize the values of the fracture mechanics parameters and the techniques used to measure them in tabular form, and this has been done in Tables 3.1 to 3.5*. However, extensive as these tables may appear, they are only representative of the many studies that have been reported in the literature; it has not been possible to include all of the data that has been published.

Fracture energy. Table 3.1 presents the surface energies, or fracture energies, that have been obtained on various cementitious systems. Most researchers have estimated

---

*The abbreviations used in these tables are given in the Appendix.

Table 3.1.   Fracture energies of cementitious systems

Investigator	Type of test	Specimen size (mm) $l \times d \times w$	Material	Fracture surface energy (J/m^2)	Remarks
Brunauer [75]	Heat of solution	–	C–S–H ('tobermorite gel')	0.386 ± 0.02	For Ca(OH)$_2$, $\gamma = 1.18$ J/m^2 For amorphous silica, $\gamma = 0.129$ J/m^2
Kantro et al [76]	Heat of solution	–	C–S–H	0.450	
Wittmann [77]	Estimated from swelling experiments	cylindrical specimens, $l = 60$, $d = 11.3$	hcp	$\gamma = 1.75 @ w/c$ $= 0.3$ $= 1.37 @ w/c$ $= 0.45$ $= 0.66 @ w/c$ $= 0.6$	Calculations based on a combination of the Griffith and Bangham equations
Kaplan [52]	SEN beams in both 3-point and 4-point bending	$406 \times 76 \times 76$ and $508 \times 152 \times 152$	mortar	1.61 (calculated) 19.3 (obtained from $G_c = 2\gamma$)	Crack assumed to pass smoothly through both cement and sand, with no interfacial failure
Moavenzadeh and Kuguel [78]	SEN beam in 3-point bending	$305 \times 25 \times 25$	hcp: 3 days 7 days 14 days 28 days mortar: 3 days 7 days 7 days concrete: 14 days 28 days	3.5 3.9 4.7 4.2 4.2 4.7 3.2 3.2 3.5	$\gamma = \dfrac{\text{measured input energy}}{2 \times \text{fracture surface area}}$ For cement and mortar, these values were twice as high as those estimated from $\gamma = G_c/2$; for concrete, the values were about 3/4 as high.
Cooper and Figg [79]	beams with three triangular notches in bending	$200 \times 25 \times 50$	hcp	dry = 14.9 wet = 12.4	γ estimated from area under load-deflection curve
Harris et al [80]	SEN beam in 3-point bending	$l = 4d$; dimension not given	frc (both glass and steel fibres)	concrete: 20–40 steel frc: 3000–4500 glass frc: 370–800	γ decreased with increasing notch/depth ratio. It was generally somewhat higher for wet specimens than for dry ones

Reference	Test method	Specimen dimensions	Material	Value	Remarks
Auskern and Horn [81]	SEN beam in 3-point bending	76 × 13 × 13	hcp polymer impregnated hcp	9.1 75	increase in fracture energy entirely due to due to the polymer phase
Aleszka and Beaumont [82]	SEN beam in 3-point bending	125 × 25 × 25	PIC fibre reinforced PIC	2,000–20,000	work of fracture calculated from the area area under the load-deflection curve
Mizutani [45]	SEN beam in tension	279 × 6.4 × 32	hcp	1.3–2.8	the fracture surface energy was higher for wet specimens than for dry ones
Ohigashi [83]	DEN beam in 3-point bending and SEN beam in 3-point bending	$l = 135$	grc mortar	~ 10,000 6–18	fracture energy for mortar decreased with increasing notch depth
Watson [84]	SEN beam in 3-point bending	250 × 20 × 20 125 × 20 × 20	hcp	2.3–3.8	the fracture energy decreased with increasing notch/depth ratio and span/depth ratio
Andonian et al [85]	SEN beam in 3-point bending	80 × 20 × 8	cellulose frc	180–970	work of fracture increased with increasing fibre content
Brandt [59]	SEN beam in 4-point bending	300 × 50 × 50	steel frc	800–1200	fracture energy increased with increasing notch depth
Chtchhourouv [86]	Compression	not given	hcp	3.2–4.2	
Mai et al [64]	SEN beam in 3-point bending	100–800 × 25 – 200 × 7.5	asbestos cement	800–3000	work of fracture decreased with increasing notch/depth ratio
Morita [87]	from COD tests	–	–	–	work of the fracture independent of notch/depth ratio
Okada et al [88]	SEN-beam in bending	390 × 100 × 47	concrete		crack depth proportional to amount of energy dissipated in fracture
Sugama et al [89]	SEN beam in 3-point bending	75 × 12.5 × 12.5	polymer concrete	0.3–4.0	
Birchall et al [48]	SEN beam in 3-point bending	not given	hcp	19–30	fracture energy increased when larger voids were removed

Table 3.1. *(continued)*

Investigator	Type of test	Specimen size (mm) $l \times d \times w$	Material	Fracture surface energy (J/m^2)	Remarks
Meiser and Tressler [90]	SEN beam in 3 point bending work-of-fracture	56 × 25 × 16	low density aluminous cement/ perlite composite	0·2–1.4 3–10	fracture energy increased with density
Sia et al [91]	beam in bending	100 × 6–10 × 25	expoxy/cement composite	40–880	

the fracture energy from tests on notched beams in bending, but as may be seen, several other techniques have also been used. The range in values is very large, particularly when data on frc or PIC are included. This in in large part due to the difficulty in estimating the true fracture surface area, because of the crack tortuosity and the extensive amount of branch cracking that occurs. Thus, for both hardened cement paste and mortar, measured fracture energies tend to be about an order of magnitude higher than the energies that would be expected from surface energy measurements of cement [75, 76] and aggregate. It may also be seen that the introduction of either fibres or polymers greatly increases the fracture energy, since both materials tend to 'tie' cracks together.

K_c *and* G_c. Most of the research on the fracture of cement and concrete has been based on the assumption that LEFM could be applied to these systems, and thus most studies have been devoted to measurements of K_c or G_c. Again, many different techniques have been used, but by far the commonest test has been the single edge notch (SEN) beam in either 3-point or 4-point bending. As may be seen from Table 3.2, many different specimen sizes have been used, and there is a fair degree of scatter in the reported results. (Note that in Table 3.2, the results are presented in essentially chronological order).

In addition to the SEN beam, however, many other techniques have also been used, as indicated in Table 3.3. Again, there is a large degree of variability. In looking at all of the data in Tables 3.2 and 3.3, it is not clear that any one technique or specimen geometry is preferable for carrying out these measurements.

The results summarized in Table 3.2 and 3.3 are also inconsistent in certain aspects. Perhaps the most troubling problem is the contradictory evidence regarding the relationship between K_c (or G_c) and the crack length, as shown in Figure 3.9. If LEFM is to apply to these materials, then the fracture parameters should be largely independent of the crack length, or the relative notch depth. And, indeed, a number of investigators [24, 40, 58, 59, 98, 99, 101, 108] have found K_c and G_c to be independent of crack length for hcp and concrete. However, Gjorv et al [40] found that while this was true for hcp, for mortar and concrete K_c decreased with increasing notch depth.

Other investigations have produced very different results. Ohigashi [83] and Harris et al [80] found that the fracture energy decreased markedly with increasing notch depth, while Mai [117] found that the work of fracture increased with increasing crack growth. Several other studies [43, 46, 122] have shown that K_c and G_c decreased with increasing notch depth, though Hillemeier and Hilsdorf [43] found that the results stabilized beyond a notch/depth ratio of 0.6. In addition, Watson [122] showed that G_c decreased as the beam span/depth ratio increased.

Still other studies have produced completely opposite results, indicating an increase in K_c or G_c with increasing relative lengths [96, 84, 152, 155]. Finally, some investigators have suggested that the fracture toughness depends not on the relative notch depth, but on the overall specimen size. In general, K_c has been found to increase as the specimen size increases [41, 55, 112, 121, 150]. To complicate matters further, where comparative tests have been carried out using different specimen geometries, different K_c values have been determined [96, 99, 112, 150].

Table 3.2. K_c and G_c from tests of single edge notched beams in bending

Investigator	Specimen dimensions (mm) $l \times d \times w$	3-Pt. Bend	4-Pt. Bend	Material	K_c MN m$^{-3/2}$	G_c N/m	Remarks
Kaplan [52]	406 × 76 × 76	✓		mortar	0.62–0.69	13.7–17.2	No correction made for subcritical crack growth. Centre-point loading gave results approximately 20% higher than third-point loading.
				concrete	0.54–0.86	10.2–11.9	
	508 × 152 × 152		✓	mortar	0.73–0.87	18.9–27.2	
				concrete	0.67–1.13	15.4–33.9	
Glucklich [71]	1067 × 102 × 51		✓	mortar		17.5–20	G_c about 7.5% lower in fatigue than in static tests.
Lott and Kesler [92]	305 × 102 × 102		✓	mortar	0.3–0.33		K_c independent of w/c ratio, increased with coarse aggregate content.
				concrete	0.34–0.40		
Moavenzadeh and Kuguel [78]	305 × 25 × 25	✓		cement	0.13–0.17	3.5–4.7	G_c, K_c largely independent of age.
				mortar	0.13–0.15	3.9–4.3	
				concrete	0.23–0.26	8.4–9.6	
Naus and Lott [53]	305 × 102 × 102		✓	cement	0.3–0.4		K_c increased with: decreasing w/c ratio, decreasing air content, increasing age, increasing maximum aggregate size, increasing volume.
	356 × 51 × 51			mortar	0.2–0.5		
				concrete	0.4–0.8		
Welch and Haisman [93]	not given			concrete	0.66–0.94	10.5–37.5	values depend on method of calculation.
George [94]	266 × 76 × 76	✓		soil	0.12–0.2	4.9–20.4	G_c independent of notch depth; increases slightly with an increase in the loading rate
				cement			
Koyanagi and Sakai [95]	388 × 100 × 47	✓		mortar	0.2–0.4	7.9–14.8	G_c, K_c increased with increasing aggregate volume for normal concrete, decreased with increasing aggregate volume for light weight concrete.
				concrete	0.35–0.5	11.8–14.8	
Brown [96]	250 × 38 × 38		✓	hcp	0.48		crack growth determined using compliance techniques. For mortar K_c increased with increasing crack growth
				mortar			

Reference	Dimensions		Material	K	G	Comments
Harris et al [80]	$l/d = 4.0$; dimensions not given	✓	frc	0.3–0.4 0.6–0.8 0.4–0.8		plain concrete K_c slightly higher for wet than for dry specimens steel frc glass frc
Okada and Koyanagi [97]	$360 \times 100 \times 47$	✓	concrete mortar	0.3–0.5 0.2–0.4	10–15 8–16	G_c and K_c increased with increasing aggregate content for normal concrete, but decreased for light weight concrete.
Walsh [61]	254–1270 long, 76–381 deep 75 mm wide	✓	concrete	0.5–1.0		Minimum beam depth for valid K_c measurements $\cong$ 230 mm.
Brown [98]	$250 \times 38 \times 38$	✓	glass frc	0.8–2.0		compliance technique used to determine crack length. K_c for gfrc increased with crack growth and with increasing fibre content.
			plain concrete	0.4		
Brown and Pomeroy [99]	$250 \times 38 \times 38$	✓	concrete hcp	0.5–1.0 0.3		K_c increased with crack growth
Togawa et al [100]	$400 \times 100 \times 100$		mortar concrete	0.2–0.4 0.3–0.5	5–16	strength of the coarse aggregate more important than the strength of the cement-aggregate interface
Yokomichi et al [101]	$450 \times 150 \times 130$	✓	mortar concrete		27–37 16–25	little influence of notch depth
Nadeau et al [102]	$228 \times 13 \times 76$	✓	hcp	0.32		good agreement with DT tests
Glucklich and Korin [103]	$228 \times 25 \times 11.5$	✓	mortar		20–40	G_c increased on drying
Aleszka and Beaumont [82]	$125 \times 25 \times 25$	✓	PIC polymer impregnated frc	0.4–1.6	5–60	both polymers and fibres increased the work to fracture

Table 3.2. *(continued)*

Investigator	Specimen dimensions (mm) $l \times d \times w$	3-Pt. Bend	4-Pt. Bend	Material	K_c MN m$^{-3/2}$	G_c N/m	Remarks
Higgins and Bailey [41]	70 × 14 × 25 to 550 × 110 × 25	✓		cement	0.3–0.7		K_c independent of notch depth, increased increasing rate of loading, increased with increasing beam depth.
Mindess and Nadeau [104]	203.2 × 50.8 × 45–254	✓		mortar concrete	0.47 0.75		K_c independent of the length of the crack front
Walsh [62]	450–5400 × 200–2400 × 40	✓		concrete	0.4–0.8		for valid tests, the minimum specimen depth must be 230 mm
Gjorv et al [40]	550 × 50 × 50	✓		hcp mortar concrete	0.1 0.15–0.2 0.13–0.24		K_c varied with notch depth for mortar and concrete, but not hardened cement paste
Gustaffson [105]	250 × 50 × 50	✓		lightweight concrete	0.3–1.2		K_c linearly related to density
Zhen [106]	275 × 30 × 25	✓		mortar	0.4–1.2		
Khrapkov et al [107]	1300 × 500 × 200	✓		concrete	0.15–0.61		K_c increased as the maximum aggregate size went from 20 to 40 mm.
Mazars [108]	1000 × 150 × 100	✓		concrete		10.2	G_c independent of test method and crack geometry
Mindess et al [44]	356 × 76 × 76		✓	cement plain concr. frc (steel and glass)	0.5–0.66 0.88 0.7–1.3	8.8–15.4 17.3 11.3–43.4	K_c, G_c insensitive to fibre intent *content*, varied with crack depth
Strange [109]	60 × 12 × 6 to 1000 × 200 × 100	✓		concrete frc	0.5–1.1		K_c increased with increasing specimen size and with increasing aggregate size. Fibres had no effect on K_c. Slow crack growth noted.

Reference	Specimen dimensions		Material		Value		Remarks
Beaumont and Aleszka [25]	125 × 25 × 25	✓	frc / PMMA impregnated frc		0.5–1.5		PMMA increases the fracture toughness
Cook and Crookham [26]	500 × 100 × 100		polymer concretes	✓	0.2–0.4		K_c increased with crack growth to a limiting value at a crack length of 35–40 mm.
Dikovskii [110]	310 × 100 × 100	✓	concrete		0.3		K_c independent of notch dimensions
Morita and Kato [111]	420 × 100 × 100	✓	concrete		0.5–0.6		
Nishioka et al [112]	530 or 260 × 60 × 60 / 530 or 260 × 150 × 150	✓	steel frc	✓	1–2		slow crack growth neglected. K_c for frc was 2–3 times greater than for plain concrete.
Swartz et al [74]	406 × 102 × 76	✓	concrete		0.1–1.6		it was concluded that K_c is not a fundamental material property
Alford and Poole [113]	152 × 25 × 25	✓	concrete with different aggregates			5–10	G_c increased with increasing 'angularity' of the aggregates
Djabarov et al [114]	160 × 40 × 40	✓	steel frc			12–35	steel fibres improved the fracture toughness
Jujii et al [115]	200 × 150 × 75 or 50	✓	mortar		$K_{IC} \simeq 0.55$		K_{IC} independent of specimen width, but K_{IIC} decreased by 30% for the wider specimens.
Kayyali et al [116]	200 × 100 × 100		hcp	✓		1–2.5	
Mai [117]	190 × 50 × 6 / 190 × 26 × 6	✓	asbestos cement		1.5–3.6		K_c in general agreement with DCB values
Modeer [118]	440 × 50 × 10 / 800 × 100 × 100		hcp / concrete			9 / 114	G_c varied with specimen depth

Table 3.2. *(continued)*

Investigator	Specimen dimensions (mm) $l \times d \times w$	3-Pt. Bend	4-Pt. Bend	Material	K_c MN m$^{-3/2}$	G_c N/m	Remarks
Morita and Kato [119]	420 × 100 × 100		✓	lightweight aggregate concrete	0.5–0.6		K_c independent of notch/depth ratio about 30% lower for lightweight concrete than for ordinary concrete.
Hornain et al [120]	40 × 13.3 × 53.3	✓		hcp / mortar / hcp/limestone interface	0.2–0.5 / 0.1–0.6 / 0.1–0.2		
Strange and Byant [55]	48 × 12 × 12 up to 1000 × 200 × 100	✓		hcp / mortar / concrete	0.6–0.8 / 0.6–0.8 / 0.7–1.1		K increased with increasing specimen size
Strange and Byant [121]	48 × 12 × 12 up to 800 × 200 × 100	✓		concrete with different aggregate sizes	0.6–1.1		K_c increased with increasing specimen size with increasing maximum size of aggregate
Watson [122]	250 × 20 × 20	✓		hcp		10–20	G_c increased with decreasing span/depth ratio
Brandt [59]	300 × 50 × 50		✓	steel frc	0.3–0.7	15–30	a great deal of scatter in the data
Kim et al [58]	292 × 76 × 76 / 292 × 38 × 76	✓		mortar	0.52 / 0.39		correcting for slow crack growth (by compliance) raised these values by 30–40%
Loland and Gjorv [123]	1500 × 100 × 100	✓		concrete		140–170	no clear relationship between G_c and tensile strength
Loland and Hustad [124]	1100 × 100 × 100	✓		concrete with silica fume		100–180	
Knox [125]	not given	✓		concrete		~18	

Reference	Dimensions	Material	✓				Comments
Petersson [126]	600 × 50 × 50	concrete	✓			60–100	G_c not affected by maximum aggregate size
Swamy [33]	500 × 100 × 100	frc, with steel, glass, polypropylene fibres	✓	1–10			the apparent K_c increases with increasing crack growth
Velazco et al [127]	457 × 76 × 38	mortar steel frc	✓	0.5 0.6–2.4			K_c depends on initial notch depth, and increases with increasing crack length
Carpinteri [128]	600 × 150 × 150	mortar concrete	✓	0.65 0.53–0.75	188 57–64		K_c increased with increasing notch/depth ratio up to about 0.3, then become stable
Huang [129]	381 × 102 × 16 610 × 203 × 13	concrete	✓	0.5–1.3 0.5–1.7			
Mindess and Bentur [60]	400 × 100 × 15–19	wood fibre frc	✓	1.4–1.7			LEFM probably not a valid approach
Nadeau et al [130]	100 × 37 × 12.7	cement mortar	✓	0.35			K_c increased from zero to 0.35 MNm$^{-3/2}$ at 40 days
Panasyuk et al [131]	160 × 40 × 40	mortar cement		0.12–0.75 0.25–1.30			
Sia et al [91]	100 × 6–10 × 25	epoxy-resin modified mortars	✓			43–80	G_c increased with epoxy content
		cellulose-cement composites			730–880		
Kitagawa and Suyama [132]	320 × 80 × 40 240 × 60 × 30 160 × 40 × 20	mortar concrete	✓	0.27 0.34			
Tait and Keenliside [133]		concrete organic frc	✓	—			K_c independent of notch/depth ratio

Table 3.3 K_c and G_c from other test geometries

Investigator	Type of test	Specimen size (mm) $l \times d \times w$	Material	K_c MNm$^{-3/2}$	G_c (N/m)	Remarks
Brown [96]	DCB	350 × 100 × 50	hcp	0.48		For mortar, K_c increased with increasing crack length
Brown and Pomeroy [99]	DCB	350 × 100 × 50	hcp mortar	0.45–0.75		The addition of aggregate increased K_c, and resulted in a progressive increase in K_c with crack growth
Mai [117]	grooved DCB	190 × 76 × 76	asbestos cement	1.7–3.7		K_c ~ independent of notch/depth ratio.
Sok et al [134]	DCB	2800 × 1100 × 300	concrete	1.6–3.6		damage zone up to 500 mm in size ahead of crack tip
Visalvanich and Naaman [135]	contoured DCB	610 × 51 × 305–127	mortar asbestos cement steel frc	1.3 4.9 1–15		K_c appears to depend on crack length; for mortar, it stabilizes only after ~ 200 mm of crack growth
Wecharatana and Shah [136]	contoured DCB	610 × 51 × 305–127	mortar concrete		~ 100 ~ 160	length of microcracked region ~ 150–200 mm. K_c reached a constant value only after ~ 1000 mm of crack growth
Chhuy et al [137]	DCB	3500 × 300 × 1100	concrete prestressed concrete	~ 3		
Neerhoff [138]	DCB	145 × 38 × 40	hcp	0.5–0.6		small changes in K_c were observed with changes in zeta-potential
Visalvanich and Naaman [139]	contoured DCB	610 × 51 × 305–127	mortar asbestos cement	1.3 4.9		need crack growth > 200 mm for valid K_c measurements
Carmichael and Jerram [140]	CT	25.4 mm thick	mortar	0.32–0.48		K_c independent of crack length

Reference	Test	Dimensions	Material	Value	Value	Comments
Patterson and Chan [34]	CT	102 × d × 102	high alumina cement with E-glass fibres		10–20	G_c appeared to be independent of specimen size
Hillemeier and Hilsdorf [43]	wedge-loaded CT	100 × 80 × 40	hcp / cement-aggregate interface	0.28–0.35 / 0.10–0.24		K_c decreases with increasing crack length, levelling off at a notch/depth ratio of ~0.6.
Hillemeier [141]	wedge-loaded CT	150 × 100 × 40	hcp	0.3–0.4		
Lenain and Bunsell [142]	CT	120 × 50 × 6.3	asbestos cement	1.7		polypropylene fibres did not change K_c
Barr et al [143]	CT	100 × d × 100	polypropylene frc	0.6–0.7		
Batson [144]	WOL	254 × 102 × 244	concrete / steel frc		60–300	G_c increased with increasing fibre content
Mindess et al [145]	DT	228 × 13 × 76	hcp	0.31–0.37		Similar to results from SEN beams
Nadeau et al [102]	DT	228 × 13 × 76	hcp	0.3–0.4		good agreement with SEN beam tests
Evans et al [24]	DT	105 × 6.3 × 57	mortar / polymer impreganted mortar	0.4–0.6 / 2.0–2.5		K_c independent of crack length beyond a crack length of 20 mm.
Wecharatana and Shah [136]	DT	813 × 38 × 152	mortar / concrete		~100 / ~160	
Wecharatana and Shah [146]	DT	813 × 38 × 152	mortar	1.3	~98	unstable crack propagation occurred at a crack length of 305–380 mm
Yam and Mindess [147]	DT	1219 × 51 × 406	hcp / steel frc	0.75 / 0.8–2.7		K_c increased with increasing fibre content
Yarema and Krestin [148]	Diametrical compression with centre slit	(l × d) 190 and 34 × 58	mortar	0.23		

Table 3.3.   (continued)

Investigator	Type of test	Specimen size (mm) $l \times d \times w$	Material	K_c MNm$^{-3/2}$	G_c (N/m)	Remarks
Lamkin and Paschenko [149]	Diametral compression with centre slit	$(l \times d)$ 100 × 150 300 × 700	concrete	0.7		
Kishitani and Maeda [63]	Diametral compression with centre slit	not given	high alumina cement mortar		8–14	for crack lengths less than 15 mm, G_c decreased
Kitagawa et al [150]	Diametral compression with centre slit	$(l \times d)$ 100 × 33 150 × 50	concrete	0.27–0.36		K_c increased with increasing specimen diameter
Pak et al [151]	diametral compression with centre slit	$(l \times d)$ 400 × 400	concrete	0.2–1.0		for LEFM to apply, the crack length must be greater than 2 times the maximum aggregate size
Romualdi and Batson [152]	plate with central slit in tension	813 × 610 × 64	concrete		5.3–12.3	G_c increased with increasing crack length
Kesler et al [46]	plate with central slit in tension	457–914 × 305 × 51	hcp mortar concrete	0.1–0.2 0.4–1.2 0.6–1.4		K_c varied considerably with crack size
Mazars [108]	plate with central cracks and single-edge cracks	600 × 340 × 80	concrete		10.2	G_c independent of crack geometry and test method
Barr and Bear [42]	CNRBB	$(l \times d)$ 150 × 41 or 54	mortar	0.3–1.0		K_c increased with increasing notch depth and with increasing rate of loading
Barr and Bear [153]	CNRBB, loaded eccentrically	$l \times 54$	mortar	0.6–1.0		K_c decreased with increasing notch depth

Reference	Specimen	$(l \times d)$	Material			Remarks
Javan and Dury [155]	CNRBB	200×100	steel and polyproylene frc	0.1–0.5		K_c increased with increasing notch depth
Yokomichi et al [101]	plate with slit in compression	$200 \times 100 \times W$	mortar		~500	G_c in compression is about 15 times as high as measured for SEN beams in bending
Desayi [155]	prism with inclined slit in compression	$305 \times 102 \times 102$	mortar concrete	0.38–0.45 0.23–0.45	50–350 50–500	concrete notch insensitive for notches less than 13 mm in length
Chtchourov [86]	compression	not given	hcp hcp at 80°C			
Barr et al [143]	cube with slits eccentrically loaded in compression	100 mm cube	polypropylene frc	0.5–0.6		similar to values obtained with CT specimens
Higgins and Bailey [41]	DEN specimen in tension	$150 \times d \times 10$ or 50 $75 \times d \times 25$	hcp mortar	0.39–0.49		K_c increased with increasing specimen size
Mizutani [45]	tension ring with a central ligament	$279 \times 6.4 \times 31.8$	hcp	0.2–0.3		

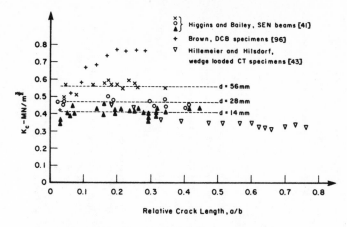

Figure 3.9.   Effect of notch depth and specimen size on K_c.

The reasons for these discrepancies are not really understood. Apart from the problem of specimen size (which will be discussed later), there are a number of reasons for this behaviour. Some of the problems are inherent in the experimental techniques themselves. Most investigators did not take into account any slow crack growth that might have occurred, thereby underestimating the effective critical crack length. When an attempt was made to consider slow crack growth, this was largely on the basis of compliance measurements. Unfortunately, the compliance of a specimen with a cast or sawn notch is not the same as that of a 'naturally' cracked specimen; in the first case there is no interaction between the sides of the crack, while in the second case, due to tortuosity and surface roughness, there may be considerable interaction between the two sides of the crack. Thus, as was shown by Kim et al [58], the compliance method for estimating method for estimating slow crack growth is inaccurate, particularly since a considerable amount of crack branching may take place near the crack tip.

In addition, the tests described above have been carried out on a wide variety of testing machines and loading frames, with very different stiffness or compliance. Though data obtained by Mindess and Bentur [156] indicate that machine stiffness has no significant effect on K_c, where fracture parameters (such as fracture energy) are obtained from the complete load-deflection curve, then machine stiffness becomes very important.

Slow crack growth parameters. In addition to the direct determination of fracture toughness, a number of studies have been carried out in an attempt to obtain sub-critical crack growth data, by determining the relationship between K_I at the crack tip and the crack velocity, V. This data, generally presented in the form of a log V vs. log K_I plot, can be defined by the slope (n) and intercept (A) of the linear portion of this plot; much of the data available is summarized in Table 3.4. The first study of this type, by Nadeau et al [102], showed that for hcp the $V–K_I$ relationship could be expressed as

$$\log V = 36 \log K_I - 200 \tag{3.3}$$

95

Table 3.4.   The slope (n) of the log V vs. log K_I plot

Investigator	Type of test	Specimen size (mm) $l \times d \times w$	Material	n	Remarks
Nadeau et al [102]	double torsion	229 × 13 × 76	hcp (saturated)	36	crack velocities measured in the range 10^{-6} to 10^{-2} m/s
Mindess et al [145]	double torsion	229 × 13 × 76	hcp (saturated) hcp (dry) hcp (low pressure steam cured) hcp (high pressure steam cured)	75 64 69 46	crack velocities measured in the range 10^{-6} to 10^{-2} m/s
Evans et al [24]	double torsion	105 × 6 × 57	mortar polymer impregnated mortar	~30	neither w/c ratio, age, nor polymer impregnation had any effect on n
Wecharatana and Shah [146]	double torsion	813 × 38 × 152	mortar	~20–30	unstable crack propagation occurred at crack lengths of about 305–380 mm
Yam and Mindess [147]	double torsion	1219 × 51 × 406	hcp glass frc steel frc	36.5 22–50 11–86	the greatest effect on n appeared to be the degree of composition, rather than fibre type of content. Fibre additions up to 2.0% by volume did not appear to restrain crack growth significantly
Mindess and Nadeau [37]	variable rate of loading on beams in 4-point bending	228 × 10 × 10	hcp mortar	17.7 14.9	these values were not consistent with values reported in refs. 102, 145.
Zech and Wittmann [157]	variable rate of loading on beams in 3-point bending	200 × 20 × 20	mortar	20–25	loading rates in the range 0.5 N/mm².sec to 30,000 N/mm².sec.
Mihashi and Wittmann [158]	variable rate of loading on beams in 3-point bending, and compression	160 × 40 × 40 and 300 × 100 × 100	mortar lightweight concrete normal concrete	18–22 25–27 25.6 58.7 26	bending compression bending compression compression

Evans et al [24] obtained comparable plots for mortar. Subsequent tests on much larger double torsion specimens [136, 146, 147] also produced similar values of n.

However, these results are not consistent with other data. In principle, the crack growth parameter n can also be obtained by determining the effect of strain rate (or stress rate) on strength σ_r, from the expression

$$\log \sigma_r = \frac{1}{n+1} \log \dot{\epsilon} + B \qquad (3.4)$$

where B is a constant. Unfortunately, Mindess and Nadeau [37] found that n-values obtained by these two methods were not in good agreement, and other studies of this type [57, 58] have also revealed wide differences in n-values. The reasons for these discrepancies are not understood.

Non-linear fracture criteria. As the inconsistencies in fracture parameters obtained by applying LEFM mounted, a number of investigators began to question the applicability of LEFM to cementitious systems. As a result, a number of non-linear (elastic-plastic) fracture criteria have been applied, primarily the J-integral, R-curve analysis, COD, and the 'fictitious' crack model (FCM). The results of tests using the first three of these are summarized in Table 3.5; the results using the fictitious crack model are described in detail elsewhere in this volume.

A number of studies have suggested that the J-integral (i.e. J_c) might be used as a suitable fracture criterion for both plain and fibre concretes. However, the variability in determinations of J_c is very large. In addition, Velazco et al [127] found that J_c was not independent of crack length. Moreover, Sih [166] has recently pointed out that, theoretically, the J-integral simply cannot be applied to composite systems, particularly systems such as frc. Thus, this approach does not now appear to be a very promising one.

R-curve analysis has also been used by a number of investigators [127, 134, 136, 163, 164]. While some studies have shown the R-curve to be independent of the test configuration [136, 163], other studies (164) have shown that R-curves are influenced by different specimen geometries. In addition, the R-curves appear to be influenced by the crack opening, with larger openings leading to higher strain energy release rates [136]. However, there is still no agreement as to whether R-curves are truly independent of specimen geometry and testing conditions, and further work in this area is probably warranted.

A few investigators have used COD to characterize the fracture parameters of fibre reinforced concrete [127, 135, 139, 165]. In general, COD values also seem to depend on specimen geometry, and do not appear to provide a useful critical value for these materials. Visalvanich and Naaman [139] have recently shown that unstable cracking appeared to occur at a critical crack opening angle of 0.13° for their large DCB specimens, and this sort of failure criterion may prove to be useful.

Finally, Hillerborg and his colleagues [118, 125, 170–175] have proposed a tied-crack model (the fictitious crack model), in which stresses are assumed to act across a crack as long as it is only narrowly opened; the region just behind the crack tip corresponds to a microcracked zone with some remaining ligaments for stress transfer. Basically, the fracture energy is measured from the test of a SEN beam in

Table 3.5. Non-linear fracture mechanics criteria

Investigator	Type of test	Specimen size (mm) $l \times d \times w$	Material	Parameter measured	Value obtained	Remarks
Mindess et al [44]	SEN beam in 4-point bending	356 × 76 × 76	hcp concrete steel frc	J_c	11–15 N/m 40–43 N/m	a high degree of variability in J_c
Mai [117]	SEN beam in 3-point bending	190 × 26 × 6	asbestos cement	J_c	340–2000 N/m	J_c agrees quite well with the unit fracture energy
Rokugo [159]	SEN beam in 4-point bending	381 × 76 × 76 381 × 152 × 76 508 × 152 × 102	mortar concrete steel frc	J_c	17–35 N/m 25 N/m 88–440 N/m	J_c independent of specimen dimensions
Brandt [59]	SEN beam in 4-point bending	300 × 50 × 50	steel frc	J_c	30–60 N/m	a great deal of scatter in J_c determinations
Carrato [160] Halvorsen	SEN beam in 4-point bending	381 × 76 × 76 381 × 152 × 76	concrete steel frc	J_c	14–26 N/m 31–315 N/m	J_c independent of specimen size and type of test
Halvorsen [161]	SEN beam in 4-point bending	381 × 76 × 76	steel frc	J_c	50–3800 N/m	J_c subject to considerable experimental variability
Halvorsen [162]	SEN beam in 4-point bending	381 × 76 × 76	hcp concrete	J_c	11 N/m 25–37 N/m	J_c subject to considerable experimental variability
Velazco et al [127]	SEN beam in 4-point bending	457 × 76 × 38	steel frc mortar	J_c	30–800 N/m 15 N/m	J_c varies greatly with initial notch depth
Carpinteri [66]	SEN beam in 3-point bending	600 × 150 × 150	concrete mortar	J_c	57 N/m 225 N/m	
Sok et al [134]	DCB	2800 × 300 × 1100	concrete	R-curve	$K_R = 3.6$ MNm$^{-3/2}$	energy necessary for crack propagation increased as the crack grew
Foote et al [163]	notch bend DCB CT	800 × 200 × W 360 × 80 × W 240 × 200 × W	asbestos cement	R-curve	$K_R = 4$–9 MNm$^{-3/2}$	K_R curves reasonably independent of specimen geometry

	Specimen type	Dimensions	Material	Parameter	Value	Remarks
Shah [164]	DT contoured DCB	813 × 38 × 152 610 × 51 × 305−127	mortar	R-curve	$G_R = 100-110$ N/m	R-curves not influenced by different specimen geometries
Velaaco et al [127]	SEN beam in 4-point bending	457 × 76 × 38	steel frc	R-curve	$K_R = 1-20$ MNm$^{-3/2}$	R-curve independent of specimen geometry
Wecharatana and Shah [136]	DT DCB	813 × 38 × 152 610 × 51 × 305−127	mortar concrete	R-curve	$G_R \simeq 100$ N/m $G_R \simeq 160$ N/m	R-curve independent of test configuration
Mai et al [64]	SEN beam in 4-point bending	100 × 25 × 7.5 to 800 × 200 × 7.5	asbestos cement	R-curve	$K_i = 1.3$ MNm$^{-3/2}$	crack extension before the plateau is 70−80 mm
Saeki and Takada [165]	plate with slit in tension	300 × 150 × 50	hcp mortar concrete	COD$_{crit}$	3−4 μm 10−12 μm 13−16 μm	
Velazco et al [127]	SEN beam in 4-point bending	457 × 76 × 38	steel frc mortar	COD$_{crit}$	40−50 μm 18 μm	COD values at peak load depend upon the notch depth
Visalvanich and Naaman [139]	contoured DCB		concrete asbestos cement steel frc	COD	critical crack opening angle = 0.13°	

a 3-point bending, as (area under the total load deformation curve + energy supplied by the weight of the beam)/(area of the ligament over the notch). This model is best suited for numerical calculations by using finite element methods. It is asserted that this model is a better representation of the fracture process than LEFM, or other non-linear critiera, but this is still a matter under active discussion.

3.7 Valid specimen size

Some of the reasons for the large experimental variability (or even inconsistency) in the fracture parameters described above have been discussed in Section 3.6. However, an even more fundamental problem than those already mentioned is that of defining a specimen size which is large enough to provide a valid test. That is, the specimen should be dimensioned so that it is large compared to the size of the micro-cracked region ahead of the crack tip. As may be seen from Table 3.6, this micro-cracked region may be very extensive, with estimates ranging from 30 to 500 mm. In addition, various tests have suggested that valid fracture parameters can only be measured after crack extensions ranging from 75 mm to 1000 mm, depending on the investigation. Though these estimates vary widely, they all imply the necessity for rather large specimens.

Various estimates of minimum specimen sizes are also given in Table 3.6. The first person explicitly to study this problem appears to have been Walsh [61, 62] who concluded that for concrete beams, the minimum specimen depth should be at least 230 mm. Most other estimates are even larger. Using the ASTM standard for plane-strain fracture toughness of metallic materials [168], which requires a specimen depth $\geqslant 5.0 (K_{IC}/\sigma_{\text{yield}})^2$, leads to a minimum specimen depth for concrete beams of about 500 mm. Carpinteri [66], from his definition of the best brittleness number, s, (see Section 4), concluded that for a concrete beam to be notch sensitive (i.e. for LEFM to be applicable), a minimum specimen depth of about 650 mm would be required. Modeer [118], using the fictitious crack model, argued that for valid fracture tests, beams should have a depth greater than 10 l_{ch}, where l_{ch} is defined as $G_C E/\sigma_t^2$, with σ_t being the tensile strength. This implies a minimum beam depth of 2–3 m for concrete. And finally, Bažant [169] has suggested that the minimum beam depth should be at least 100 times the size of the microcracked region, or about 500 times the maximum aggregate size; for 20 mm aggregate, this would lead to a minimum beam depth of 10 m!

It is clear from the above that the minimum beam depth should be quite large, probably at least 1 m. However, to the best of the present author's knowledge, no tests on concrete beams of this size have ever been carried out, though a few large DT and DCB specimens have been tested. Thus, it is almost certain that very few, if any, truly valid fracture mechanics tests have ever been carried out on concrete (though some valid tests have probably been carried out on hcp). It is therefore not surprising that there is so little agreement in the values of the fracture parameters that have been obtained.

Table 3.6. Minimum specimen sizes for valid fracture mechanics tests

Investigator	minimum specimen size (mm) $l \times d \times w$			Size of microcracked region ahead of crack tip (mm)	Remarks
	hcp	mortar	concrete		
Lenain and Bunsell [142]				~30	from tests on CT specimens of asbestos cement
Entov and Yagust [167]			✓	~80–100	
Sok et al [134]			✓	~500	from tests on DCB specimen
Chhuy et al [137]			✓	~150–200	K_c reached a constant value only after crack propagation of ~1000 mm
Visalvanich and Naaman [135, 139]		✓			from tests on DCB specimens, need a crack extensions of 75 mm in asbestos cement and ~200 mm in mortar for valid K_c values
Wecharatana and Shah [136, 146]		✓	✓		unstable crack propagation occurred at a crack length of 305–380 mm for double torsion tests
ASTM E-399-81 [168]	~480 × 120 × 60 ~150 × 144 × 60	~800 × 200 × 100 ~250 × 240 × 100	~2000 × 500 × 250 ~625 × 600 × 250		SEN beam in bending Compact tension specimen
Walsh [61]			$d > 150$		SEN beam in bending
Walsh [62]			$d > 230$		SEN beam in bending
Bazant [169]	$d \sim 10$	$d \sim 2500$	$d \sim 10{,}000!$		cross-sectional dimensions > 100 times the size of the microcracked zone, or about 500 times the maximum aggregate size
Modeer [118]	$d \sim 50$–100	$d \sim 1000$–2000	$d \sim 2000$-3000		SEN beam in bending
Carpinteri [66]	$d \sim 100$	$d \sim 250$	$d \sim 650$		SEN beam in bending

3.8 Conclusions

In spite of all of the work that has been carried out to determine the fracture mechanics parameters for concrete, there is still no agreement as to which parameters best characterize the fracture of concrete. Moreover, because most tests have been carried out on specimens which were too small, there have probably been very few valid fracture mechanics tests carried out on concrete. There is still disagreement on how to analyze the test data that have been obtained. Clearly, a great deal of further work is warranted. Only when these problems are solved will fracture mechanics become a useful tool for the analysis of concrete structures.

References

[1]   Richart, F.E., Brandtzaeg, A. and Brown, R.L., A study of the failure of concrete under combined compressive stresses, *Bulletin No. 185*, Engineering Experiment Station, University of Illinois (1928).

[2]   Griffith, A.A., The phenomena of rupture and flow in solids, *Philosophical Transactions, Royal Society of London, Series A221*, pp. 163–198 (1920).

[3]   Mindess, S., The cracking and fracture of concrete: an annotated bibliography, 1928–1981, *Fracture Mechanics of Concrete*, edited by F.H. Wittmann, Elsevier Scientific Publishing Company, (in press) (1983).

[4]   Berger, R.L., Calcium hydroxide: its role in the fracture of tricalcium silicate paste, *Science*, 175, 4022, pp. 626–629 (1972).

[5]   Berger, R.L., Lawrence, F.V. and Young, J.F., Studies on the hydration of tricalcium silicate pastes II. Strength development and fracture characteristics, *Cement and Concrete Research*, 3, 5, pp. 497–508 (1973).

[6]   Berger, R.L., personal communication.

[7]   Young, J.F., personal communication.

[8]   Walsh, D., Otooni, M.A., Taylor, M.E. and Marcinkowski, M.J., Study of portland cement fracture surfaces by scanning electron microscopy techniques, *Journal of Materials Science*, 9, 3, pp. 423–429 (1974).

[9]   Marchese, B., SEM topography of twin fracture surfaces of alite pastes 3 years old, *Cement and Concrete Research*, 7, 1, pp. 9–18 (1977).

[10] Jennings, H.M. personal communication.

[11] Diamond, S. and Mindess, S., Scanning electron microscopic observations of cracking in portland cement paste, *Proceedings of the Seventh International Congress on the Chemistry of Cement*, Paris, Vol. III, pp. VI. 114–119 (1980).

[12] Mindess, S. and Diamond, S., A device for direct observation of cracking of cement paste or mortar under compressive loading within a scanning electron microscope, *Cement and Concrete Research*, 12, 5, pp. 569–576 (1982).

[13] Diamond, S., Mindess, S. and Lovell, J., Use of a Robinson backscatter detector and 'wet cell' for examination of wet cement paste and mortar specimens under load, *Cement and Concrete Research*, in press (1982).

[14] Zaitsev, J. and Wittmann, F.H., Simulation of crack propagation and failure of concrete, *Matériaux et Constructions*, 14, 83, pp. 357–365 (1981).

[15] Dantu, P., Etude des contraintes dans les milieux hétérogenes. Application au beton, *Annales de L'Institut Technique de Batimen et des Travaux Publics*, 11, 121, pp. 55–77 (1958).

[16] Swamy, R.N., Aggregate-matrix interaction in concrete systems, *Structure, Solid Mechanics and Engineering Design*, edited by M. Te'eni, Proceedings of the Southampton 1969 Civil Engineering Materials Conference, Wiley-Interscience, pp. 301–315 (1971).

[17] McReath, D.R., Newman, J.B. and Newman, K., The influence of aggregate particles on the local strain distribution and fracture mechanism of cement paste during drying shrinkage and loading to failure, *Matériaux et Constructions*, 2, 7, pp. 73–84 (1969).

[18] Hsu, T.T.C., Slate, F.O., Sturman, G.M. and Winter, G., Microcracking of plain concrete and the shape of the stress-strain curve, *Journal of the American Concrete Institute*, 60, 2, pp. 209–224 (1963).

[19] Swamy, R.N., Fracture phenomena of hardened paste, mortar and concrete, *Proceedings of the International Conference on Mechanical Behaviour of Materials, Kyoto, 1971*, The Society of Materials Science, Japan, Vol. IV, pp. 132–142 (1972).

[20] Evans, R.H. and Marathe, M.S., Microcracking and stress-strain curves for concrete in tension, *Matériaux et Constructions*, 1, 1, pp. 61–64 (1968).

[21] Wang, P.T., Shah, S.P. and Naaman, A.E., Stress-strain curves of normal and lightweight concrete in compression, *Journal of the American Concrete Institute*, 75, 11, pp. 603–611 (1978).

[22] Mindess, S. and Diamond, S., A preliminary SEM study of crack propagation in mortar, *Cement and Concrete Research*, 10, 4, pp. 509–519 (1980).

[23] Mindess, S. and Diamond, S. The cracking and fracture of mortar, *Matériaux et Constructions*, 15, 86, pp. 107–113 (1982).

[24] Evans, A.G., Clifton, J.R. and Anderson, E., The fracture mechanics of mortars, *Cement and Concrete Research*, 6, 4, pp. 535–548 (1976).

[25] Beaumont, P.W.R. and Aleszka, J.C., Cracking and toughening of concrete and polymer-concrete dispersed with short steel wires, *Journal of Materials Science*, 13, 8, pp. 1749–1760 (1978).

[26] Cook, D.J. and Crookham, G.D., Fracture toughness measurements of polymer concretes, *Magazine of Concrete Research*, 30, 105, pp. 205–214 (1978).

[27] Munoz-Escalona, A. and Ramos, C., Fracture morphology and mechanical properties of thermocatalytically polymerized MMA-impregnated mortar, *Journal of Materials Science*, 13, 2, pp. 301–310 (1978).

[28] Bhargava, J. and Rehnstrom, A., High speed photography for fracture studies of concrete, *Cement and Concrete Research*, 5, 3, pp. 239–248 (1975).

[29] Romualdi, J.P. and Mandel, J.A., Tensile strength of concrete affected by uniformly distributed and closely spaced short lengths of wire reinforcement, *Journal of the American Concrete Institute*, 61, 6, pp. 657–670 (1964).

[30] Romualdi, J.P., The static cracking stress and fatigue strength of concrete reinforced with short pieces of thin steel wire, *The Structure of Concrete*, edited by A.E. Brooks and K. Newman, Proceedings of an International Conference, London, 1965. Cement and Concrete Association, London, pp. 190–201 (1968).

[31] Parimi, S.R. and Rao, J.J.S., Effectiveness of random fibres in fibre-reinforced concrete, *Proceedings of the International Conference on Mechanical Behaviour of Materials, Kyoto, 1971*, The Society of Materials Science, Japan, Vol. V, pp. 176–186 (1972).

[32] Aleszka, J. and Schnittgrund, G., An evaluation of the fracture of plain concrete, fibrous concrete, and mortar using the scanning electron microscope, *Technical Report M-122*, Construction Engineering Laboratory, Champaign, Illinois (1975).

[33] Swamy, R.N., Influence of slow crack growth on the fracture resistance of fibre cement composites, *International Journal of Cement Composites*, 2, 1, pp. 43–53 (1980).

[34] Patterson, W.A., and Chan, H.C., Fracture toughness of glass fibre-reinforced cement, *Composites*, 6, 1, pp. 102–104 (1975).

[35] Rüsch, H., Researches toward a general flexural theory for structural concrete, *Journal of the American Concrete Institute*, 57, 1, pp. 1–28 (1960).

[36] McHenry, D. and Shideler, J.J., Review of data on effect of speed in mechanical testing of concrete, *Symposium on Speed of Testing of Nonmetallic Materials*, ASTM STP 185, ASTM, Philadelphia, Pa., pp. 72–82 (1955).

[37] Mindess, S. and Nadeau, J.S., Effect of loading rate on the flexural strength of cement and mortar, *American Ceramic Society Bulletin*, 56, 4, pp. 429–430 (1977).

[38] Ziegeldorf, S., Müller, H.S. and Hilsdorf, H.K., A model law for the notch sensitivity of brittle materials, *Cement and Concrete Research*, 10, 5, pp. 589–599 (1980).

[39] Shah, S.P. and McGarry, F.J., Griffith fracture criterion and concrete, *Journal of the Engineering Mechanics Division, ASCE*, 97, EM6, pp. 1663–1676 (1971).

[40] Gjorv, O.E., Sorensen, S.I. and Arnesen, A., Notch sensitivity and fracture toughness of concrete, *Cement and Concrete Research*, 7, 3, pp. 333–344 (1977).

[41] Higgins, D.D. and Bailey, J.E., Fracture measurements on cement paste, *Journal of Materials Science*, 11, 11, pp. 1995–2003 (1976).

[42] Barr, B. and Bear, T., A simple test of fracture toughness, *Concrete*, 10, 4, pp. 25–27 (1976).

[43] Hillemeier, B. and Hilsdorf, H.K., Fracture mechanics studies on concrete compounds, *Cement and Concrete Research*, 7, 5, pp. 523–536 (1977).

[44] Mindess, S., Lawrence, F.V. and Kesler, C.E., The J-integral as a fracture criterion for fibre reinforced concrete, *Cement and Concrete Research*, 7, 6, pp. 731–742 (1977).

[45] Mizutani, K., An indirect tension test with applications to fracture of brittle materials, Ph.D. Thesis, Purdue University, West Lafayette, Indiana (1978).

[46] Kesler, C.E., Naus, D.J. and Lott, J.L., Fracture mechanics – its applicability to concrete. *Proceedings of the International Conference on Mechanical Behaviour of Materials, Kyoto, 1971*, The Society of Materials Science, Japan, Vol. IV, pp. 113–124 (1972).

[47] Saouma, V.E., Ingraffea, A.R. and Catalano, D.M. Fracture toughness of concrete – K_{IC} revisited, *Journal of the Engineering Mechanics Division, ASCE* (in press).

[48] Birchall, J.D., Howard, A.J. and Kendall, K., Flexural strength and porosity of cements, *Nature*, 289, 5796, pp. 388–390 (1981).

[49] Wright, W. and Byrne, J.G., Stress concentration in concrete, *Nature*, 203 4952, pp. 1374–1375 (1964).

[50] Evans, R.H. and Marathe, M.S., Stress distribution around holes in concrete, *Matériaux et Constructions*, 1, 1, pp. 57–60 (1968).

[51] Imbert, I.D.C., The effect of holes on tensile deformation in plain concrete, *Highway Research Record No. 324*, Highway Research Board, Washington, D.C., pp. 54–65 (1970).

[52] Kaplan, M.F., Crack propagation and the fracture of concrete, *Journal of the American Concrete Institute*, 58, 5, pp. 591–610 (1961).

[53] Naus, D.J. and Lott, J.L., Fracture toughness of portland cement concretes, *Journal of the American Concrete Institute*, 66, 6, pp. 481–489 (1969).

[54] Kamiyama, S., An effect of notch for uni-axial tensile strength of mortar and concrete, *Review of the Twentieth General Meeting*, The Cement Association of Japan, Tokyo, pp. 153–156 (1966).

[55] Strange, P.C. and Bryant, A.G., Experimental tests on concrete fracture, *Journal of the Engineering Mechanics Division, ASCE*, 105, EM2, pp. 337–343 (1979).

[56] Cook, D.J. and Crookham, G., A discussion of the paper 'Notch sensitivity and fracture toughness of concrete, by O.E. Gjorv, S.I. Sorensen and A. Arnesen, *Cement and Concrete Research*, 8, 3, pp. 387–388 (1978).

[57] Cook, D.J. and Crookham, G., Fracture toughness measurements of polymer concrete, *Magazine of Concrete Research*, 30, 105, pp. 205–214 (1978).

[58] Kim, M.M., Ko, H.-Y. and Gerstle, K.H., Determination of fracture toughness of concrete, *Fracture in Concrete*, edited by W.F. Chen and E.C. Ting, American Society of Civil Engineers, New York, pp. 1–14 (1980).

[59] Brandt, A.M., Crack propagation energy in steel fibre reinforced concrete, *International Journal of Cement Composites*, 2, 1, pp. 35–42 (1980).

[60] Mindess, S. and Bentur, A., The fracture of wood fibre reinforced cement, *International Journal of Cement Composites and Lightweight Concrete*, 4, 4, (in press) (1982).

[61] Walsh, P.F., Fracture of plain concrete, *Indian Concrete Journal*, 44, 11, pp. 469–470, 476 (1972).

[62] Walsh, P.F., Crack initiation in plain concrete, *Magazine of Concrete Research*, 28, 94, pp. 37–41 (1976);

[63] Kishitani, K. and Maeda, K., Compressive fracture of cracked mortar, *Review of the Thirtieth General Meeting*, The Cement Association of Japan, Tokyo, pp. 216–217 (1976).

[64] Mai, Y.W., Foote, R.M.L. and Cotterell, B., Size effects and scaling laws of fracture in asbestos cement, *International Journal of Cement Composites*, 2, 1, pp. 23–34 (1980).

[65] Halvorsen, G.T., Toughness of portland cement concrete, Ph.D. Thesis, University of Illinois, Urbana (1979).

[66] Carpinteri, A., Static and energetic fracture parameters for rocks and concretes, *Matériaux et Constructions*, 14, 81, pp. 151–162 (1981).

[67] Neville, A.M., Some aspects of the strength of concrete, *Civil Engineering (London)*, Part I: 54, 639, pp. 1153–1156 (1969); Part II: 54, 640, pp. 1308–1310 (1959); Part III: 54, 641, pp. 1435–1439 (1959).

[68] Mindess, S., The application of fracture mechanics to cement and concrete: A historical review, *Fracture Mechanics of Concrete*, edited by F.H. Wittmann, Elsevier Scientific Publishing Company, in press (1983).

[69] Mindess, S., The fracture of fibre reinforced and polymer impregnated concretes: a review, *Fracture Mechanics of Concrete*, edited by F.H. Wittmann, Elsevier Scientific Publishing Company, in press (1983).

[70] Glucklich, J., Fracture of plain concrete, *Journal of the Engineering Mechanics Division, ASCE*, 89, EM6 pp. 127–138 (1983).

[71] Glucklich, J., Static and fatigue fractures of portland cement mortar in flexure, *Proceedings of the First International Conference on Fracture, Japan, 1965*, The Japanese Society for Strength and Fracture of Materials, Vol. 3, pp. 1343–1382 (1966).

[72] Glucklich, J., The effect of microcracking on time-dependent deformation and the long-term strength of concrete, *The Structure of Concrete*, edited by A.E. Brooks and K. Newman, Proceedings of an International Conference, London, 1965, Cement and Concrete Association, London, pp. 176–189 (1968).

[73] Glucklich, J., The strength of concrete as a composite material, *Proceedings of the International Conference on Mechanical Behaviour of Materials, Kyoto, 1971*, The Society of Materials Science, Japan, Vol. IV, pp. 104–112 (1972).

[74] Swartz, S.E., Hu, K.-K. and Jones, G.L., Compliance monitoring of crack growth in concrete, *Journal of the Engineering Mechanics Division, ASCE*, 104, EM4, pp. 789–800 (1978).

[75] Brunauer, S., Tobermorite gel – the heart of concrete, *American Scientist*, 50, 1, pp. 210–229 (1962).

[76] Kantro, D.L., Weise, C.H. and Brunauer, S., Paste hydration of betadicalcium silicate, tricalcium silicate, and alite, *Symposium on Structure of Portland Cement Paste and Concrete, Special Report 90, Highway Research Board, Washington, D.C., pp. 309–327 (1966).*

[77] Wittmann, F.H., Surface tension, shrinkage and strength of hardened cement paste, *Matériaux et Constructions*, 1, 6, pp. 547–552 (1968).

[78] Moavenzadeh, F. and Kuguel, R., Fracture of concrete, *Journal of Materials*, 4, 3, pp. 497–519 (1969).

[79] Cooper, G.A. and Figg, J., 1. Fracture studies of set cement paste, *Journal of the British Ceramic Society*, 71, pp. 1–4 (1972).

[80] Harris, B., Varlow, J. and Ellis, C.D., The fracture behaviour of fibre reinforced concrete, *Cement and Concrete Research*, 2, 4, pp. 447–461 (1972).

[81] Auskern, A. and Horn, W., Fracture energy and strength of polymer impregnated cement, *Cement and Concrete Research*, 4, 5, pp. 785–795 (1974).

[82] Aleszka, J.C. and Beaumont, P.W.R., The work of fracture of concrete and polymer impregnated concrete composites, *Proceedings, First International Congress on Polymers in Concrete, 1971*, The Construction Press, Lancaster, England, pp. 269–275 (1976).

[83] Ohigashi, T., Measurement of effective fracture energy of glass fibre reinforced cement, *Testing and Test Methods of Fibre Cement Composites, RILEM Symposium 1978*, The Construction Press, Lancaster, England, pp. 67–78 (1978).

[84] Watson, K.L., The estimation of fracture surface energy as a measure of the 'toughness' of hardened cement paste, *Cement and Concrete Research*, 8, 5, pp. 651–656 (1978).

[85] Andonian, R., Mai, Y.W. and Cotterell, B., Strength and fracture properties of cellulose fibre reinforced cement composites, *International Journal of Cement Composites*, 1, 4, pp. 151–158 (1979).

[86]   Chtchourov, A.F., Microtexture et resistance mecanique du cement durci, *Proceedings of the Seventh International Congress on the Chemistry of Cement*, Paris, Vol. III, pp. 404–410 (1980).

[87]   Morita, K., Evaluation of fracture toughness by means of the work of fracture, *The 35th Annual Meeting of the Civil Engineering Institute of Japan*, Vol. V, pp. 269–270 (1980).

[88]   Okada, K., Koyanagi, W. and Rokugo, K., Energy approach to flexural fracture process of concrete, *Transaction of the Japan Society of Civil Engineers*, 11, pp. 301–304 (1980).

[89]   Sugama, T., Kukacka, L.E. and Horn, W., The effect of β-C_2S/Class H cement mixed fillers on the kinetics and mechanical properties of polymer concrete, *Cement and Concrete Research*, 10, 3, pp. 413–424 (1980).

[90]   Meiser, M.D. and Tressler, R.E., Mechanical properties of a low density aluminous cement/perlite composite, *American Ceramic Society Bulletin*, 60, 9, pp. 901–905 (1981).

[91]   Sia, T.K., Mai, Y.W. and Cotterell, B., Strength and fracture properties of epoxy-cement composites, *Second Australian Conference on Engineering Materials, Sydney*, pp. 515–529 (1981).

[92]   Lott, J. and Kesler, C.E., Crack propagation in plain concrete, *Symposium on Structure of Portland Cement Paste and Concrete*, Special Report 90, Highway Research Board, Washington, D.C., pp. 204–218 (1966).

[93]   Welch, G.B. and Haisman, B., The application of fracture mechanics to concrete and the measurement of fracture toughness, *Matériaux et Constructions*, 2, 9, pp. 171–177 (1969).

[94]   George, K.P., Theory of brittle fracture applied to soil cement, *Journal of the Soil Mechanics and Foundation Division, ASCE*, 96, SM3, pp. 991–1010 (1970).

[95]   Koyanagi, W. and Sakai, K., Observations on the crack propagation process of mortar and concrete, *Review of the Twenty-Fifth General Meeting*, The Cement Association of Japan, Tokyo, pp. 153–157 (1971).

[96]   Brown, J.H., Measuring the fracture toughness of cement paste and mortar, *Magazine of Concrete Research*, 24, 81, pp. 185–196 (1972).

[97]   Okada, K. and Koyanagi, W., Effect of aggregate on the fracture process of concrete, *Proceedings of the International Conference on Mechanical Behaviour of Materials, Kyoto, 1971*, The Society of Materials Science, Japan, Vol. IV, pp. 72–83 (1972).

[98]   Brown, J.H., The failure of glass-fibre-reinforced notched beams in flexure, *Magazine of Concrete Research*, 25, 82, pp. 31–38 (1973).

[99]   Brown, J.H. and Pomeroy, C.D., Fracture toughness of cement paste and mortars, *Cement and Concrete Research*, 3, 4, pp. 475–480 (1973).

[100]   Togawa, K., Satoh, T. and Araki, K., Parameters on the fracture toughness of mortar and concrete, *Review of the Twenty-Seventh General Meeting*, The Cement Association of Japan, pp. 117–120 (1973).

[101]   Yokomichi, H., Fujita, Y. and Saeki, N., Experimental researches on crack propagation of plain concrete, *Review of the Twenty-Seventh General Meeting*, The Cement Association of Japan, pp. 144–147 (1973).

[102]   Nadeau, J.S., Mindess, S. and Hay, J.M., Slow crack growth in cement paste, *Journal of the American Ceramic Society*, 57, 2, pp. 51–54 (1974).

[103]   Glucklich, J. and Korin, U., Effect of moisture content on strength and strain energy release rate of cement mortar, *Journal of the American Ceramic Society*, 58, 11–12, pp. 517–521 (1975).

[104]   Mindess, S. and Nadeau, J.S., Effect of notch width on K_{IC} for mortar and concrete, *Cement and Concrete Research*, 6, 4, pp. 529–534 (1976).

[105]   Gustaffson, P.J., Brottmakaniska studier; lattbetong och Fiber-armerad betong, *Rapport TVBM-5001*, Department of Building Materials, The Lund Institute of Technology Lund, Sweden (1977).

[106]   Zhen, L., Fracture surface energy and interface bond in concrete, *International Conference on Bond in Concrete*, Paisley, Scotland, 1982; Supplementary Papers, 26–39 (1982).

[107]   Khrapkov, A.A., Trapesnikov, L.P., Geinats, G.S., Paschenko, V.I. and Pak, A.P., The application of fracture mechanics to the investigation of cracking in massive concrete

construction elements of dams, *Fracture 1977*, Proceedings of the 4th International Conference on Fracture, Waterloo, Ontario. Vol. 3, pp. 1211–1217 (1977).

[108] Mazars, J., Existence of a critical strain energy release rate for concrete, *Fracture 1977*, Proceedings of the 4th International Conference on Fracture, Waterloo, Ontario, Vol. 3, pp. 1205–1209 (1977).

[109] Strange, P.C., The fracture of plain and fibre concrete, Ph.D. Thesis, University of Auckland, New Zealand (1977).

[110] Dikovskii, I.A., On fracture toughness of concrete, *Rabotosposobnost' Stroit. Materialov v Usloviakh Vozd. Razlichnykh Expluat. Fakturov. Mezhvuzovsky Sbornik*, Kazan, K. Kh. T.I. in S.M. Kirova, 1, pp. 17–18 (1978).

[111] Morita, K. and Kato, K., Fundamental study on fracture toughness and evaluation by acoustic emission technique of concrete, *Review of the Thirty-Second General Meeting*, The Cement Association of Japan, Tokyo, pp. 138–139 (1978).

[112] Nishioka, K., Yamakama, S., Hirakawa, K. and Akihama, S., Test method for the evaluation of the fracture toughness of steel fibre reinforced concrete, *Testing and Test Methods of Fibre Cement Composites*, RILEM Symposium 1978, The Construction Press, Lancaster, England, pp. 87–98 (1978).

[113] Alford, N.McN. and Poole, A.B., The effect of shape and surface texture on the fracture toughness of mortars, *Cement and Concrete Research*, 9, 5, pp. 583–589 (1979).

[114] Djabarov, N., Wehrstedt, A., Weiss, H-J. and Yamboliev, K., Investigations on the fracture behaviour of steel fibre reinforced concrete, *Proceedings of the Second National Conference on Mechanics and Technology of Composite Materials, Varna, 1979*, Sofia, Bulgaria, pp. 593–596 (1979).

[115] Jujii, K., Haga, K. and Fujii, S., A study on progressive failure of brittle materials by fracture mechanics, *Memoirs of the Faculty of Engineering*, Tokushima University, Japan, 24, pp. 1–11 (1979).

[116] Kayyali, O.A., Page, C.L. and Ritchie, A.G.B., Frost action on immature cement paste – effects on mechanical behaviour, *Journal of the American Concrete Institute*, 76, 11, pp. 1217–1225 (1979).

[117] Mai, Y.W., Strength and fracture properties of asbestos-cement mortar composites, *Journal of Materials Science*, 14, 9, pp. 2091–2102 (1979).

[118] Modeer, M., A fracture mechanics approach to failure analysis of concrete materials, *Report TVBM-1001*, Division of Building Materials, University of Lund, Sweden (1979).

[119] Morita, K., and Kata, K., Fundamental study on evaluation of fracture toughness of artificial lightweight aggregate concrete, *Review of the Thirty-Third General Meeting*, The Cement Association of Japan, Tokyo, pp. 175–177 (1979).

[120] Hornain, H., Mortureux, B. and Regourd, M., Physico-chemical and mechanical aspects of the cement paste-aggregate bond, *Liaisons Pates de Cement Matériaux Associes*, Colloque International, Toulouse, Nov. 1982, pp. C.56–C.65 (1982).

[121] Strange, P.C. and Bryant, A.H., The role of aggregate in the fracture of concrete, *Journal of Materials Science*, 14, pp. 1863–1868 (1979).

[122] Watson, K.L., Reply to a discussion of 'The estimation of fracture surface energy as a measure of the "toughness" of hardened paste', by Y.W. Mai, *Cement and Concrete Research*, 9, 4, pp. 541–544 (1979).

[123] Loland, K.E. and Gjorv, O.E., Ductility of concrete and tensile behaviour, *Research Report BML 80.613*, University of Trondheim, NTH, Division of Building Materials, Trondheim, Norway (1980).

[124] Loland, K.E. and Hustad, T., Load response of C-25 concrete with and without addition of silica fume, *Research Report STF65 A80048*, Cement and Concrete Research Institute, University of Trondheim, NTH, Trondheim, Norway (1980).

[125] Knox, W.R.A., Fracture mechanisms in plain concrete under compression, *Fracture Toughness of High Strength Materials: Theory and Practice*, Publication 20, Iron and Steel Institute, London, pp. 158–152 (1970).

[126] Petersson, P.E., Fracture energy of concrete: practical performance and experimental results, *Cement and Concrete Research*, 10, 1, pp. 91–101 (1980).

[127]   Velazco, F., Visalvanich, K. and Shah, S.P., Fracture behaviour and analysis of fibre reinforced concrete beams, *Cement and Concrete Research*, 10, 1, pp. 41–51 (1980).

[128]   Carpinteri, A., Experimental determination of fracture toughness parameters K_{IC} and J_{IC} for aggregative materials, *Advances in Fracture Research*, Proceedings of the 5th International Conference on Fracture, Cannes, 1981, Pergamon Press, Vol. 4, pp. 1491–1498 (1981).

[129]   Huang, C.-M.J., Finite element and experimental studies of stress – intensity factors for concrete beams, Ph.D. Thesis, Kansas State University, Manhattan, Kansas (1981).

[130]   Nadeau, J.S., Bennett, R. and Mindess, S., Acoustic emission in the drying of hardened cement paste and mortar, *Journal of the American Ceramic Society*, 64, 7, pp. 410–415 (1981).

[131]   Panasyuk, V.V., Berezhnitsky, L.T. and Chubrikov, V.M., Estimation of cement concrete crack resistance according to the failure viscosity, *Beton i Zhelezobeton*, Moscow, pp. 19–20 (1981).

[132]   Kitagawa, H. and Suyama, M., Fracture mechanics study on the size effect for the strength of cracked concrete materials, *Proceedings, Nineteenth Japan Congress on Materials Research*, The Society of Materials Science, Tokyo, pp. 156–159 (1976).

[133]   Tait, R.A. and Keenliside, W., Toughness of cellulose cement composites, *Advances in Fracture Research*, Proceedings of the 5th International Conference on Fracture, Cannes, 1981, Pergamon Press, Vol. 2, pp. 1091–1108 (1981).

[134]   Sok, C., Baron, J. and Francois, D., Mechanique de la rupture appliquee au beton hydraulique, *Cement and Concrete Research*, 9, 5, pp. 641–648 (1979).

[135]   Visalvanich, K. and Naaman, A.E., Evaluation of fracture techniques in cementitious composites, *Fracture in Concrete*, edited by W.F. Chen and E.C. Ting, American Society of Civil Engineers, New York, pp. 65–81 (1980).

[136]   Wecharatana, M. and Shah, S.P., Resistance to crack growth in portland cement composites, *Fracture in Concrete*, edited by W.F. Chen and E.C. Ting, American Society of Civil Engineers, New York, pp. 82–105 (1980).

[137]   Chhuy, S., Benkirane, M.E., Baron, J. and Francois, D., Crack propagation in prestressed concrete. Interaction with reinforcement, *Advances in Fracture Research*, Proceedings of the 5th International Conference on Fracture, Cannes, 1981, Pergamon Press, Vol. 4, pp. 1507–1514 (1981).

[138]   Neerhoff, A., Correlation between fracture toughness and zeta potential of cement stone, *Proceedings of the NATO Advanced Research Institute, Adhesion Problems in the Recycling of Concrete*, Saint-Remy-Les-Chevreuse, 1980, Plenum Publishing Corp., pp. 267–284 (1981).

[139]   Visalvanich, K. and Naaman, A.E., Fracture methods in cement composites, *Journal of the Engineering Mechanics Division, ASCE*, 107, pp. 1155–1171 (1981).

[140]   Carmichael, G.D.T. and Jerram, K., The application of fracture mechanics to prestressed concrete pressure vessels, *Cement and Concrete Research*, 3, 4, pp. 454–467 (1975).

[141]   Hillemeier, B., Bruchmechanische untersuchungen des rissfortschritts in zementgebundenem werkstoffen, Dr.-Ing. Thesis, Karlsruhe University (1976).

[142]   Lenain, J.C. and Bunsell, A.R., The resistance to crack growth of asbestos cement, *Journal of Materials Science*, 14, 2, pp. 321–332 (1979).

[143]   Barr, B.I.G., Evans, W.T. and Dowers, R.C., Fracture toughness of polypropylene fibre concrete, *International Journal of Cement Composites and Lightweight Concrete*, 3, pp. 115–122 (1981).

[144]   Batson, G.B., Strength of steel fibre reinforced concrete in adverse environments, *Special Report M-218*, Construction Engineering Research Laboratory, Champaign, Illinois (1977).

[145]   Mindess, S., Nadeau, J.S. and Hay, J.M., Effects of different curing conditions on slow crack growth in cement paste, *Cement and Concrete Research*, 4, 6, pp. 953–965 (1974).

[146]   Wecharatana, M. and Shah, S.P., Double torsion tests for studying slow crack growth of portland cement mortar, *Cement and Concrete Research*, 10, 6, pp. 833–844 (1980).

[147]   Yam, A.S.-T. and Mindess, S., The effects of fibre reinforcement on crack propagation in concrete, *International Journal of Cement Composites and Lightweight Concrete*, 4, 2, pp. 83–93 (1982).

[148] Yarema, S.Ya. and Krestin, G.S., Determination of the modulus of cohesion of brittle materials by compressive tests on disc specimens containing cracks, *Fiziko-Khimicheskaya Mekhanika Materialov*, 2, pp. 10–14 (1966).

[149] Lamkin, M.S. and Paschenko, V.I., Determination of the critical stress intensity factor of concrete, *Isvestiya VNIGG im B.E. Vedeneeva, Leningrad*, 99, pp. 234–239 (1972).

[150] Kitagawa, H., Kim, S. and Suyama, M., Determination of fracture toughness of concrete materials by diametral compression tests, *Proceedings, Nineteenth Japan Congress on Materials Research*, The Society of Materials Science, Tokyo, pp. 160–163 (1976).

[151] Pak, A.P., Trapesnikov, L.P., Sherstobitova, T.P. and Yakovleva, E.N., Experimental and analytical determination of the critical length of crack in concrete, *Izvestiya VNIIG im. B.E. Vedeneeva*, 116, pp. 50–54 (1977).

[152] Romualdi, J.P. and Batson, G.B., Mechanics of crack arrest in concrete, *Journal of the Engineering Mechanics Division, ASCE*, 89, EM6, pp. 775–790 (1963).

[153] Barr, B. and Bear, T., Fracture toughness, *Concrete*, 11, 4, pp. 30–32 (1977).

[154] Javan, L. and Dury, B.L., Fracture toughness of fibre reinforced concrete, *Concrete*, 13, 12, pp. 31–33 (1979).

[155] Desayi, P., Fracture of concrete in compression, *Matériaux et Constructions*, 10, 57, pp. 139–144 (1977).

[156] Mindess, S. and Bentur, A., The effect of the longitudinal stiffness of the testing machine on the strength and fracture of hardened cement paste (work in progress).

[157] Zech, B. and Wittmann, F.H., Variability and mean value of strength of concrete as function of load, *Journal of the American Concrete Institute*, 77, 5, pp. 358–362 (1980).

[158] Mihashi, H. and Wittmann, F.H., Stochastic approach to study the influence of rate of loading on strength of concrete, *Heron*, 25, 3, 54 pp. (1980).

[159] Rokugo, K., Experimental evaluation of fracture toughness parameters of concrete, M.S. Thesis, Department of Civil Engineering, University of Illinois, Urbana, Illinois (1979).

[160] Carrato, J.L., Experimental evaluation of the J-integral, M.S. Thesis, Department of Civil Engineering, University of Illinois, Urbana, Illinois (1980).

[161] Halvorsen, G.T., J-integral study of steel fibre reinforced concrete, *International Journal of Cement Composites*, 2, 1, pp. 13–22 (1980).

[162] Halvorsen, G.T., J_m toughness comparison for some plain concretes, *International Journal of Cement Composites*, 2, 3, pp. 143–148 (1980).

[163] Foote, R.M.L., Cotterell, B. and Mai, Y.W., Crack growth resistance curve for a cement composite, *Advances in Cement-Matrix Composites*, Proceedings, Materials Research Society, Symposium L, Boston, pp. 135–144 (1980).

[164] Shah, S.P., Fracture in fibre reinforced concrete, *Advances in Cement-Matrix Composites*, Proceedings, Materials Research Society, Symposium L, Boston, pp. 83–89 (1980).

[165] Saeki, N. and Takada, N., Crack propagation and COD of concrete, *The 35th Annual Meeting of the Civil Engineering Institute of Japan*, V, pp. 267–268 (1980).

[166] Sih, G.C., Personal communication.

[167] Entov, V.M. and Yagust, V.I., Experimental investigation of the regularities of quasistatic development of macrocracks in concrete, *Izvestiya Akademii Nauk SSR, Mekhanika Tverdogo Tela*, 4, pp. 93–103 (1975).

[168] ASTM E-399-81, Test for plane-strain fracture toughness of metallic materials, *Annual Book of ASTM Standards*, American Society of Testing and Materials, Philadelphia, 10, (1981).

[169] Bazant, Z.P., Materials behaviour under various types of loading, *High Strength Concrete*, edited by S.P. Shah, Proceedings of a Workshop held at the University of Illinois at Chicago Circle, December 1979, pp. 79–92 (1980).

[170] Hillerborg, A., Modeér, M. and Petersson, P.-E., Analysis of crack formation and crack growth in concrete by means of fracture mechanics and finite elements, *Cement and Concrete Research*, 6, 6, pp. 773–782 (1976).

[171] Hillerborg, A., A model for fracture analysis, *Report TVBM-3005*, Division of Building Materials, Lund Institute of Technology. Lund, Sweden (1978).

[172] Hillerborg, A., Analysis of fracture by means of the fictitious crack model, particularly for fibre reinforced concrete, *International Journal of Cement Composites*, 2, 4, pp. 177–184 (1980).

[173] Petersson, P.E., Fracture mechanical calculations and tests for fibre-reinforced cementi-
       tious materials, *Advances in Cement-Matrix Composites*, Proceedings, Materials Research
       Society, Symposium L, Boston, pp. 95–106 (1980).
[174] Petersson, P.E., Fracture energy of concrete: method of determination, *Cement and
       Concrete Research*, 10, 1, pp. 78–89 (1980).
[175] Petersson, P.E., Crack growth and development of fracture zones in plain concrete and
       similar materials, Ph.D. Thesis, Lund Institute of Technology, Sweden, 1981; *Report
       TVBM-1006*, Division of Building Materials, Lund, Sweden (1981).

Appendix: Abbreviations used in Tables 3.1 to 3.6

CT   compact tension
CNRBB   circumferentially notched round bar in bending
COD   crack opening displacement
DCB   double cantilever beam
DEN   double edge notched
DT   double torsion
frc   fibre reinforced cement (or concrete)
G_c critical strain energy release rate
G_R critical strain energy release rate obtained from R-curve analysis
hcp   hydrated cement paste
J_c critical value of the J-integral
K_c critical stress intensity factor
K_R critical stress intensity factor obtained from R-curve analysis
n slope of the log crack velocity vs. log K_I plot
SEN   single edge notched
WOL   wedge opening loaded
γ surface energy (or fracture surface energy)

Dependence of concrete fracture toughness on specimen geometry and on composition

4.1 Introduction

The field of fracture mechanics originated in the 1920's with A.A. Griffith's work on fracture of brittle materials such as glass. Its most significant applications, however, have been for controlling brittle fracture and fatigue failure of metallic structures such as pressure vessels, airplanes, ships and pipelines. Considerable development has occurred in the last twenty years in modifying Griffith's ideas or in proposing new concepts to account for the ductility typical of metals. As a result of these efforts, standard testing techniques have been available to obtain fracture mechanics parameters for metals, and design based on these parameters are included in relevant specifications.

Brittle-fracture design considerations. In a linear elastic analysis of a two-dimensional symmetrical specimen, the stresses in the region of a crack tip in a body subjected to tensile stresses normal to the plane of the crack (mode I deformation) can be expressed as

$$\sigma_x, \sigma_y = K_1 r^{-1/2} f(\theta) \tag{4.1}$$

The stress distribution is shown in Figure 4.1. Note that the function $f(\theta)$ does not depend on the specimen or the loading geometry. Consequently, it is reasonable to formulate failure criteria in terms of load and geometry dependent term K_1, the stress intensity factor.

One of the underlying principles of fracture mechanics is that unstable fracture occurs when the stress intensity factor at the tip of the crack reaches a critical value. This value is a material toughness property and also depends on loading rate and constraint as follows:

K_c = critical stress intensity factor for static loading and plane-stress conditions of variable constraint and depends on specimen thickness, crack size and the length of the uncracked ligament.

K_{1c} = critical stress intensity factor for static loading and plane-strain conditions for maximum constraint (to plastic flow). This value is minimum for thick plates

111

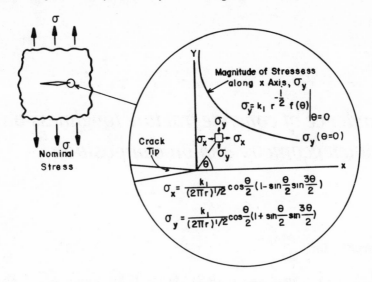

Figure 4.1.   Elastic stress distribution ahead of a crack.

K_{1d} = critical stress intensity factor for dynamic loading and plane-strain conditions where

$$K_c, K_{1c} \text{ and } K_{1d} = C\sigma\sqrt{a} \qquad (4.2)$$

C = a constant depending on specimen and crack geometry

σ = a nominal stress

a = flaw size

Each of these K values may also be a function of temperature.

By knowing the critical value of stress intensity factor for a particular thickness, loading rate and temperature, the designer can determine flaw size that can be tolerated for a given design stress level, or he can determine the design stress level that can be tolerated for an existing crack that may be present in a structure.

Experimental Determination of K_{1c}. The critical value of stress intensity factor for metallic material is a thickness dependent property. Over a certain range of thickness, the critical combination of load and crack length at instability (K_c) decreases with an increase in thickness reaching a rather constant minimum value (K_{1c}). At the crack tip prior to crack extension, a plastic zone (fracture process zone) exists (Figure 4.2). The size of this zone is much larger at the surface of sheet specimens than at the center where the stresses in the perpendicular directions (σ_z) provide constraint to plastic flow. For thick specimens, the effects of triaxial stresses (plane-strain condition) is predominant and, as a result, crack extension takes place without any appreciable plastic flow. Thus, a minimum value toughness (K_{1c}) is observed for sufficiently thick plate specimens.

If the length of the uncracked ligament is small compared to the size of the plastic zone, then elastic plane-strain condition are not possible. Some idea of the size of the plastic zone can be obtained by substituting $r = r_y$ and $\sigma_y = \sigma_{ys}$ in Equation (4.1). Thus, the radius of the plastic zone is proportional to the quantity $(K_1/\sigma_{ys})^2$.

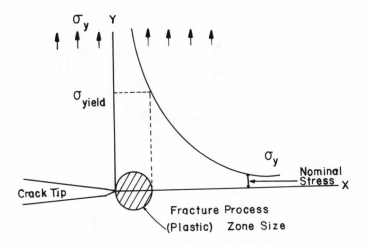

Figure 4.2.   Nonlinearity near a crack-tip.

To obtain a unique lower bound value of material fracture toughness for metals, specimens are specially designed to assure that plane-strain conditions exist during crack propagation. The test methods and the design of specimens are described in relevant ASTM specifications (for example, ASTM E-339). Two recommended test specimens are bend specimen and compact tension specimen (Figure 4.3). To assure elastic plane-strain behavior, the thickness of the specimen (B), the length of the crack (a) and the length of the uncracked ligament ($W-a$) should be substantially larger than the size of the plastic zone. This is assured by the following minimum specimen dimensions:

$$a, B \text{ and } (W-a) \geqslant 2.5 \left(\frac{K_{1c}}{\sigma_{ys}}\right)^2 \tag{4.3}$$

These specifications assure that the specimen thickness, crack length and the length of the uncracked ligament are approximately 50 times the radius of the plastic zone. Note that if plane-strain conditions are not maintained, then the value of critical stress intensity factor (K_c) will not be unique for a given material, but will depend on the test specimen dimensions.

Nonlinearity in cement based composites. Many attempts have been made to apply linear elastic fracture mechanics to cementitious composites. A comparison of values of critical stress intensity factor (K_c) obtained by various investigators is shown in Table 4.1 (taken from [1]). It can be seen that even for otherwise similar materials composition, an unacceptably large variation in K_c value is reported. There are several possible reasons for this discrepancy.

(1) For calculations of K_c, most investigators generally take the initial notch depth a_0 as the length of the crack a in equation (4.2). However, there is evidence of a substantial crack growth prior to unstable fracture. This can be seen from the load-CMD curves plotted in Figure 4.4 [1]. These results are for notched-beam specimens

113

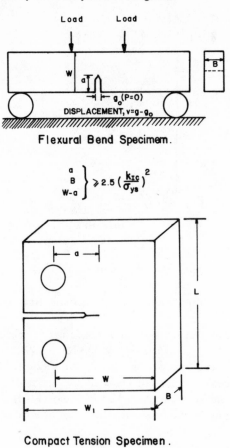

Figure 4.3.   Standard specimens for fracture toughness testing of metals.

of plain and fiber reinforced mortar specimens subjected to a four-point loading arrangement. During loading, deflection, crack mouth opening (Figure 4.5), crack extension (with an accuracy of 3×10^{-4} mm) were measured. A substantial slow crack growth occurred even for plain mortar prior to the maximum load. Unless this slow crack growth is included in the analysis, an accurate value of K_c cannot be determined. However, very few reported values include an accurate measurement of slow crack growth.

(2) Cracks in concrete are rough and tortuous. This is partly a result of cracks going around the aggregates (or fiber debonding in case of fiber reinforced concrete). This resistance to debonding during a crack extension (also called aggregate interlock) provides crack closing pressure near the tip of the crack. The extent of this crack closing pressure will depend on the size of the inclusion, bond behaviour of the inclusion in the matrix and geometry of the specimen. In addition to the traction forces behind the crack tip, there may be microcracking [2–4] ahead of the crack tip. This combined nonlinear zone around the crack-tip is referred to as the process zone. Unless the size of the process zone is small, one cannot use LEFM to predict K_c. This size of the process zone cannot be estimated from equations such as equation (4.3)

Table 4.1   Comparison of critical stress intensity factors (MNm$^{-3/2}$)

Investigators Year of reports Type of testing	Cement paste	Mortar	Concrete	Steel fibre reinforced concrete		
				$V_f = 0.5\%$	$V_f = 1\%$	$V_f = 2\%$
Nishioka, et al.						
1978 3-BD	–	–	0.73	–	1.47–1.76	1.42–1.71
4-BD	–	–	0.73	–	1.42–1.71	1.86–2.01
Mindess, et al.						
1977 4-BD	0.49–0.66	–	0.87	0.75–0.90	0.85–1.06	1.00–1.31
Gjorv et al.						
1977 3-BD	0.09–0.11	0.16–0.2	0.07–0.24	–	–	–
Hillemeir, et al.						
1977 CT	0.31	0.37	–	–	–	–
Brown, et al.						
1973 4-BD	0.3	0.65–0.9	–	–	–	–
DCB	0.43	0.60–1.1	–	–	–	–
Harris, et al.						
1972 3-BD						
Dry	–	0.4	–	–	–	0.43
Wet	–	0.41				0.57
Kesler, et al.						
1972 CCP	0.14–0.25	0.4–1.2	0.5–1.4	–	–	–
Velazco, et al.						
1980	–	0.5	–	0.6	0.9–1.3	2.4

3-BD – 3 points bending   4-BD – 4 points bending   CT = compact tension   DCB = Double cantilever beam   CCP = Center-cracked plate under tension

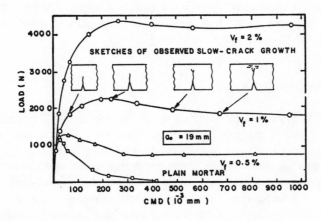

Figure 4.4.   Load-CMD plots for different fiber volume.

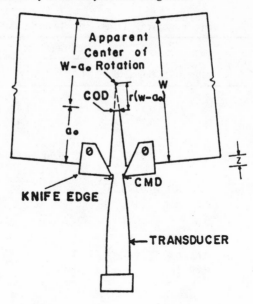

Figure 4.5.   Representation of the notch profile during bending.

because there is no plasticity in cement matrix and equation (4.3) does not include the effect of the size of the aggregates.

The effect of the size of the aggregate on fracture toughness can be seen in Figure 4.6. In this figure, results of center-notched and double-edged-notched tensile specimens made with pure cement paste and with mortar (cement + sand) are shown [5]. It appears that for comparable dimensions of test specimens and crack length, cement paste is notch-sensitive whereas mortar is not. This may be a result of the differing size of the process zone.

In this chapter some experimental and analytical approaches to modify the concepts of LEFM to include the effects of slow crack growth and crack-tip nonlinearity are described. Additional details are given in [1, 6–9].

4.2 Resistance curves

To characterize the resistance to fracture of materials during the incremental slow crack growth, R-curve analysis has been proposed [10]. R-curve can be considered as the resistance to crack propagation in terms of energy absorbed. R-curves can be obtained by plotting, for example, the strain energy release rate at each crack initiation (G_R) against the actual crack extension. Example of an R-curve is shown in Figure 4.7. It is observed that with increasing crack extension, G_R increases as a result of slow-stable crack growth. Eventually the curve may become horizontal when the fracture toughness becomes independent of the crack extension (at G_{ss}). The strain energy release rates for increasing applied loads P_1, P_2 and P_3 are shown in Figure 4.7 as dashed lines. When these lines become tangent to the resistance curve, an existing

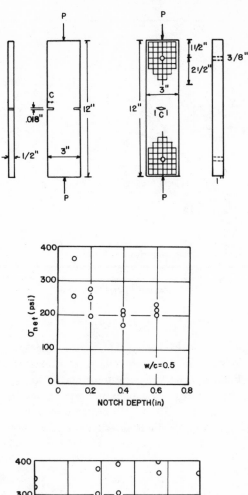

Figure 4.6.   Effects of size of aggregates on notch-sensitivity.

notch of length a_{01}, a_{02}, etc. will extend. The applied strain energy release rate at this occurrence (G_C) is a function of initial notch depth (a_0) and cannot be considered as material property. The R-curves, however, seem to be independent of the specimen geometry and thus can be considered a valid fracture parameter [1, 11, 12], provided, as shown later, the process zone is included in the analysis.

117

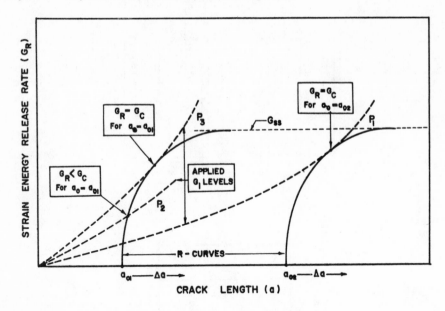

Figure 4.7.   An example of an *R*-curve.

Compliance definitions and strain energy release rate. To obtain *R*-curves, a plot of strain energy release rate against corresponding crack extension is necessary. Such curves have been obtained from the measurements of load, appropriate load-line deflection and the corresponding crack extension on notched beam, compact tension and double cantilever specimens [12–14]. It can be shown that for an elastic material and an unstable crack propagation, the strain energy release rate is given by [15]:

$$G_I = \frac{P^2}{2t_n} \frac{dC}{da} \tag{4.4}$$

where P is the load, t_n is the thickness in the plane of crack, C is the compliance of the test specimen, a is the crack length and dC/da is the rate of change of compliance (Figure 4.8a).

For a stable crack growth where the crack propagation occurs with a rising load-deflection curve, equation (4.4) can be modified as follows (Figure 4.8b):

$$G_I = \frac{P_1 P_2}{2t_n} \frac{dC_s}{da} \tag{4.5}$$

where P_1 and P_2 are two consecutive neighboring loads. Note that the value of G_I corresponds to the area of the shaded portion in Figure 4.8b. The above formula is a good approximation for a continuous curve such as shown in Figure 4.8c if the observation points are sufficiently close. The compliance measured this way is termed the secant compliance (C_s). This definition of compliance has been also called quasi-static compliance [11, 13].

The method of calculating G_I from the secant compliance measurements assumes that the material is elastic and that there will not be any irreversible deformations

118

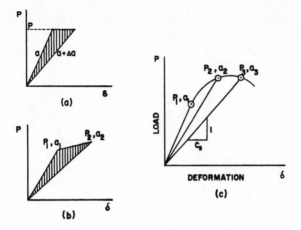

Figure 4.8.   Methods of calcuations of G_R.

upon unloading. To account for the permanent or inelastic deformations, an unloading-reloading technique has been used [12, 14, 16]. After measurements of corresponding loads and crack growth, the specimen is unloaded and then reloaded (Figure 4.9a). Instead of the secant compliance, the reloading compliance (C_R) is used for the calculation of the strain energy release rate:

$$G_I = \frac{P_1 P_2}{2t_n} \frac{dC_R}{da} \tag{4.6}$$

Although equaton (4.6) does give the change in elastic strain energy per unit crack extension, it does not include the inelastic energy absorbed during crack growth (area *OADC*, Fig. 4.9b). As a result, the strain energy release rate calculated using reloading compliance underestimates the resistance to crack growth. To include the change in strain energy due to both elastic and inelastic deformations, the definition of strain energy release rate was modified as:

$$G_R = \frac{P_1 P_2}{2t_n} \left[\frac{dC_R}{da} + \left(\frac{P_1 + P_2}{P_1 P_2} \right) \frac{d\delta_P}{da} \right] \tag{4.7}$$

where δ_P is the permanent deformation associated with the loads P_1 and P_2, dC_R/da and $d\delta_p/da$ are respectively the change of reloading compliance and permanent deformations with respect to crack extension, and G_R is the modified strain energy release rate. The modified strain energy release rate can also be expressed as:

$$G_R = \frac{P_1 P_2}{2t_n} \frac{dC_m}{da} \tag{4.8}$$

where

$$\frac{dC_m}{da} = \frac{dC_R}{da} + \left(\frac{P_1 + P_2}{P_1 P_2} \right) \frac{d\delta_p}{da} \tag{4.9}$$

Note that the strain energy corresponding to equation (4.7) is the area *OABC* in Figure 4.9b whereas for equation (4.6), it is the area *BCD*. In derivation of equation (4.7), it is assumed that the inelastic energy indicated by area *OABC* in Figure 4.9b is absorbed by the slow crack growth.

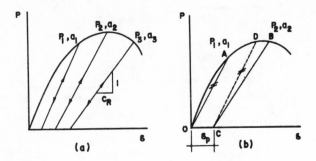

Figure 4.9.   Compliance measurements during slow crack growth.

The concept of the modified strain energy release rate and the modified compliance was developed in an attempt to apply linear elastic fracture mechanics to materials exhibiting slow crack growth and nonlinear behavior. The results of the experiments performed in this study showed that *R*-curves calculated using this approach do seem to give a valid fracture parameter. Note that to calculate G_R, both the knowledge of the rates of change per unit crack extension of reloading compliance (C_R) and the permanent deformations upon unloading (δ_P) are necessary.

4.3 Theoretical model

To include the effect of large-scale crack-tip nonlinearity and to predict the extent of this nonlinear process zone, a theoretical model has been developed [8]. The basic concepts of the model are described in this section. The analysis of crack propagation in unreinforced matrix and the modifications for fiber reinforced composites are presented.

Basic concepts and assumptions of the model. A crack initiating from a notch in Mode I opening is shown in Figure 4.10. The actual length of the crack can be divided into two zones: a traction free length *a*, and a nonlinear zone l_p where the (crack opening resisting) traction exists across the zone. If it is assumed that the stresses in the nonlinear zone are purely uniaxial, then it can be represented as a traction free length on which closing pressure $p(x)$ exists. Thus, the actual crack is replaced with an effective (elastic) crack a_{eff} such that $a_{eff} = a + l_p$, where *a* is the actual traction free length and l_p is the idealized length of the process zone (Figure 4.10). The concept is similar to that originally proposed by Dugdale [17]. The stresses and strains in a body containing the effective crack, a_{eff}, can be predicted using the concepts of LEFM. Under a given load, the crack is assumed to propagate when the crack opening displacement (COD) at the tip of the traction free length *a*, reaches a critical value equal to η_{max}

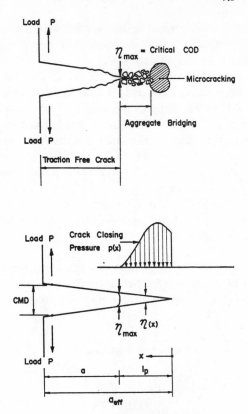

Figure 4.10.   A model of an idealized nonlinear process zone.

(Figure 4.10). The concept of critical crack opening displacement has been forwarded by other investigators [17–19]. This criteria of crack extension also provides a means of differentiating between a and l_p.

The closing pressure in the process zone depends on the crack opening displacement, and this dependence is obtained from the uniaxial tensile test as described later. Note that the value of η_{max} is also obtained from the uniaxial tensile test.

Using these concepts, the length of the process zone was calculated from the available results for three types of specimens; double cantilever, double torsion and the notched-beam specimens. Although some of the computational procedure was different for each type of specimen, the basic approach is outlined below.

Since the pressure distribution in the nonlinear zone depends on the crack opening displacement, which in turn depends on the applied load, specimen geometry, the length of the process zone and the closing pressure itself, an iterative procedure was used to calculated the length of the process zone as follows:

1.   Consider a given traction free crack length a, just prior to its further extension,
2.   Assume a crack profile and a nonlinear zone of length l_p,
3.   Knowing l_p and crack opening displacement in the process zone, calculate the closing pressure distribution ($p(\text{x})$).

121

4. For a given specimen geometry, the applied load P, and the crack closing pressure $p(x)$, calculate using the theory of elasticity, the crack opening displacements for the effective crack a_{eff},

5. If the crack opening displacement at the tip of the visible crack length a is η_{max}, then the assumed value of l_p is a correct one; otherwise another value is assumed and the above steps are repeated until that equality is satisfied,

6. For the value of l_p for which the COD at $x = l_p$ (see Figure 4.10) is equal to η_{max}, calculate the value of the crack mouth displacement (CMD, that is, COD at $x = a_{eff} = a + l_p$ (see Figure 4.10). If this value does not correspond to the experimentally measured value, then ~~charge~~ the crack profile and repeat steps 2 to 6 until it does.   *Change*

Note that since the calculation of l_p is based on the measured values of crack mouth displacement, the model is not restricted to small scale process zone as many previous ones are [12, 17–20].

Proposal model for fiber reinforced concrete. A crack just prior to its extension in Mode I opening in a fiber reinforced concrete specimen is shown in Figure 4.11. The effective crack in this case sustains two types of closing pressure; due to fiber bridging

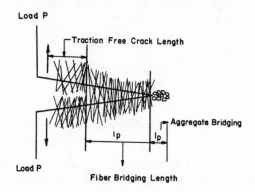

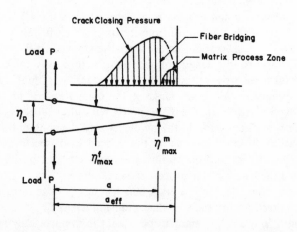

Figure 4.11.   A model of an idealized fiber-bridging zone.

and due to the matrix process zone. The traction free region and fiber bridging zone (l_f) are separated where the crack opening displacement is η^f_{max}. This value cannot exceed half the length of the fiber ($l/2$) since at that opening all the fibers are completely pulled out and the closing pressure is zero. The boundary between the fiber bridging zone and the matrix process zone is equal to η^m_{max} (or η_{max} in Figure 4.10). This value is taken as the critical COD value of the matrix. The closing pressure due to fiber bridging is assumed to be obtained from the uniaxial tests as described in the next section. Except for the addition of fiber bridging pressure, the computation of calculation of l_p, and crack mouth displacement (η_p in Figure 4.11, CMD in Figure 4.10) is essentially the same as described earlier.

4.4 Uniaxial tensile stress-displacement relationship

The proposed model depends on two materials parameters: σ-η relationship for the matrix and for the fiber reinforced composite. Experiments with plain and reinforced concrete specimens subjected to uniaxial tension in a relatively stiff testing system have shown that the post-peak displacements are essentially a result of opening of a single crack [21, 22]. For example, steel fiber reinforced mortar specimens were loaded in parallel with steel rods to obtain stable post-peak response [22]. It was observed that during the descending part, the total displacement of the specimens (regardless of the gauge length) was the same as the width of a single crack. If the microcracking occurring prior to the peak load can be ignored, then the uniaxial stress-displacement relationship in the post-peak region (σ vs. η) can be assumed to be equivalent to the closing pressure ($p(\kappa)$) vs. the crack opening displacement (η) relationship needed for the proposed theoretical model.

A compilation of some available data for fiber reinforced concrete [22–24] on stress-displacement relationship in the post-cracking stage revealed that the data can be uniquely presented when they are nondimensionalized as shown in Figure 4.12. In this figure, the closing pressure is divided by the maximum post-cracking stress whereas the crack opening displacement is divided by half the length of the fiber (η^f_{max}). A unique relationship was obtained for the data (which were all for straight, smooth steel fibers) regardless of the length, diameter and volume fraction of fibers. This relationship was expressed as:

$$\frac{\sigma}{\sigma_{max}} = \left[1 - \frac{\eta}{\eta^f_{max}} \right]^2 \tag{4.10}$$

where σ_{max} is the maximum post-cracking stress and η^f_{max} is the maximum pulled out distance of the fibers which was taken as $l/2$. Note that once σ_{max} is determined, equation (4.10) provides a simple and an accurate means of estimating the relationship between the closing pressure and the crack opening displacement for smooth steel fibers. There are many analytical and empirical expressions developed to predict σ_{max} [25].

For plain concrete, the data from Evens and Marathe [8] were used to develop analytical relationship needs for the model (4.10).

123

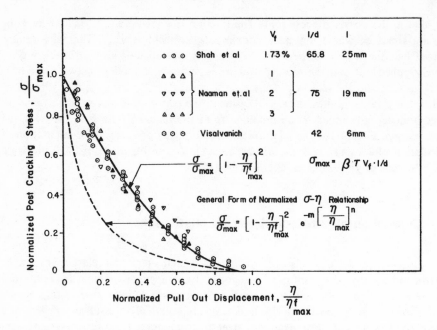

Figure 4.12. Closing pressure vs. crack opening displacement relationship.

4.5 Experimental investigation

The slow crack growth was observed by testing double torsion specimens (Figure 4.13), double cantilever beam specimens (Figure 4.14), and notched beam specimens (Figure 4.3). The detailed design of these three types of specimen is given in [1, 7, 24].

Double torsion specimens. Double torsion specimens have been recently used by many investigators to study slow crack growth since it can be shown that the rate of change of elastic compliance with crack growth is constant [26, 27]. The elastic compliance can be expressed as:

$$C = \frac{3w_m^2 a}{Wt^3 G\psi(t/W)} \tag{4.11}$$

where W, w_m and t are defined in Figure 4.13; G is the shear modulus of elasticity and ψ is the thickness correction factor [27]. By combining equations (4.11) and (4.10) it can be seen that the strain energy release rate (or the stress intensity factor) is independent of crack length. Thus, with a single double torsion specimen, it is possible to measure G_I values for several different crack lengths.

The dimensions of the double torsion specimens used in this study were 32 in. (812.8 mm) long, 6 in. (152.4 mm) wide and $1\frac{1}{2}$ in. (38.1 mm) thick. With these dimensions it was possible to study crack growth from about 3 in. (76.2 mm) to 28 in. (711.8 mm) [26]. To keep the crack growth in a predetermined plane, it was found necessary to provide a double groove along the centerline of the specimen (Figure 4.13). This premolded groove made the thickness normal to the crack front

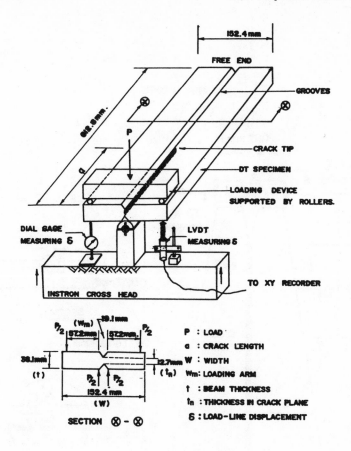

Figure 4.13.   Double torsion beam.

(t_n) equal to $\frac{1}{2}$ in. (12.7 mm). Preliminary tests showed that it was not necessary to provide an initial notch. As a result, no premolded notch was provided and the length of the first observed crack was considered as an initial crack length (a_0). Since these specimens were relatively thick, the thickness correction was applied when using equation (4.11) according to [27].

Double cantilever specimens. Double cantilever specimens can be designed so as to make the strain energy release rate (or stress intensity factor) independent of crack length. The elastic compliance of a double cantilever specimen can be expressed as [28]:

$$C = \frac{24}{Et}\left[0.33\left(\frac{a}{W}\right)^3 + 0.61\left(\frac{a}{W}\right)^2 + 0.532\left(\frac{a}{W}\right)\right] \tag{4.12}$$

where t and W are defined in Figure 4.14 and E is the Young's modulus of elasticity. The width of the specimen was tapered in such a manner that dC/da calculated from equation (4.12) was constant for the values of 'a' between 4 in. (101.6 mm) to 17 in.

125

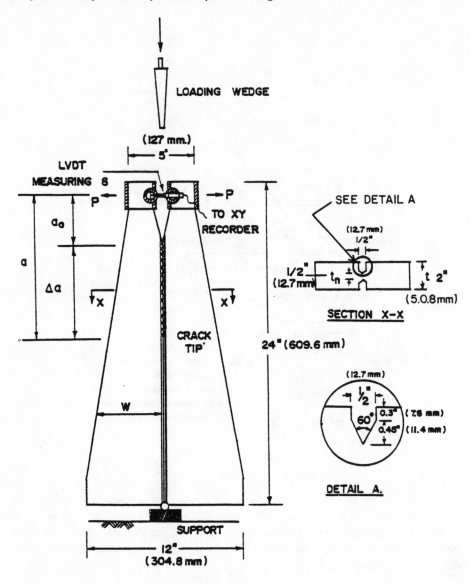

Figure 4.14.   Double cantilever beam.

(431.8 mm). The dimensions of the specimen were 24 in. (609.6 mm) long, 2 in. (50.8 mm) thick and had a tapered width from 5 in. (127 mm) to 12 in. (304.8 mm). The specimens had a 4 in. (101.6 mm) initial notch and premolded double groove which kept crack propagation along the centerline (Figure 4.14).

Notched-beam specimens. The specimens were 38 mm wide, 76 mm deep 457 mm long and tested in a four-point loading arrangement. The specimens were loaded so that the length of the central constant moment zone was 127 mm. Beams were cast with a

126

pre-molded sharp triangular notch at the center as well as without any initial notch. Four different notch depths (a_0) were used: 9.5, 19.0, 38.0 and 57.0 mm.

Testing details. Specimens were tested on an Instron testing machine. The load was applied at a constant crosshead displacement rate. The loading was stopped at frequent intervals when the crosshead displacement was held constant. During this interval, the crack growth was detected by a microscope for the double torsion and notched-beam specimens and by a telescope for the double cantilever. The accuracy of crack length measurement was 0.004 in. (0.1016 mm). After the measurement of crack length, the DT and DC specimens were unloaded and then reloaded (Figure 4.15). No unloading, reloading was done for notched-beam specimens. For these specimens, during testing a continuous record of load, cross-head displacement and CMD was made.

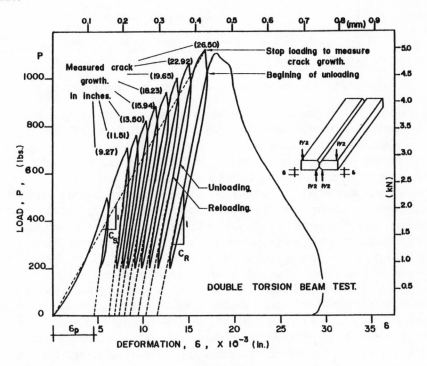

Figure 4.15. Typical load-deformation curves.

4.6. Comparison with experimental data

Process zone length. Specimens were calculated using the theoretical model described before. These values of process zone were based on a close agreement between experiments and predictions of the load-line deflection (CMD) (Figure 4.10). It was observed that during the progressive slow crack growth, the length of the process zone did not change. This length was about 3 inches (77 mm) (Figure 4.16) for DCB specimens. A similar length of the process zone can also be deduced from the data of Lott, Kesler

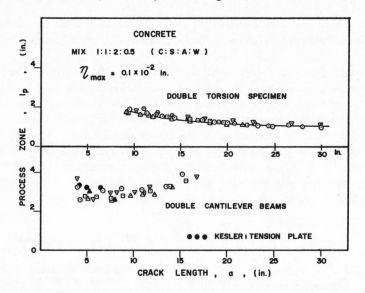

Figure 4.16.   Effect of specimen geometry on the size of the process zone.

and Naus [28]. They treated plate specimens of concrete with a central notch and wedge loading. They measured strain values at various distances from the crack-tip. If it is assumed that the crack growth occurs at the peak value of the load, and that the elastic limit tensile strain of concrete is 100 microstrain, then a value of the nonlinear zone just prior to crack growth is about 3 inches (Figure 4.16).

Lenain and Bunsell [12] measured the length of the process zone during slow crack growth for asbestos cement compact-tension specimens using acoustic emission technique. They observed that the length of the process zone at each crack initiation during testing remain essentially constant. This constant value was about 30 mm.

The length of the process zone and the critical crack opening displacement have been measured for PMMA using optical interference microscopy for compact tension specimens [30]. A constant value of l_p as well as η_{max} at crack growth were reported. Note that the length of the process zone for PMMA is of the order of 40×10^{-3} mm; 30 mm for asbestos cement specimens; and 75 mm for concrete specimens. These large differences must, among other things, be related to the size of the microstructure.

The length of the process zone was about 25 mm for DT specimens (Figure 4.16). For notched-beam specimens the length of the process zone depended on the dimension of the uncracked ligament $(W-a)$ (Figure 4.17).

Prediction of fracture resistance curve. To theoretically predict R-curve using the modified definition of strain energy release rate, one needs to know the unloading compliances and permanent deformations. The theoretical R-curves were calculated and compared for DT and DCB specimens. No experimental R-curves were available for NB specimens.

To theoretically calculate the unloading compliance and the permanent deformations, it was assumed that the displacement of the effective crack front in the nonlinear process zone are completely nonrecoverable. For example, for double

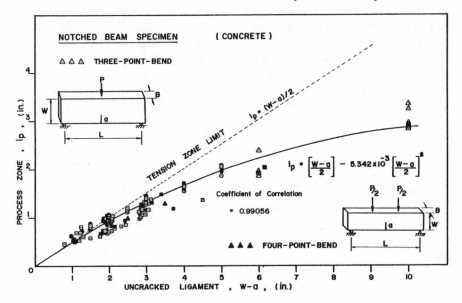

Figure 4.17.   Theoretical values of process zone for notched-beam specimens.

cantilever specimens, the unloading compliance was calculated from using elastic beam theory for a cantilever beam of length a rather than a_{eff}.

The predicted values of R-curves are compared with the measured one for a set of double-cantilever and the double torsion specimens (Figure 4.18). The comparison was found satisfactory. Note that the R-curves shown in Figure 4.18 are not identical for

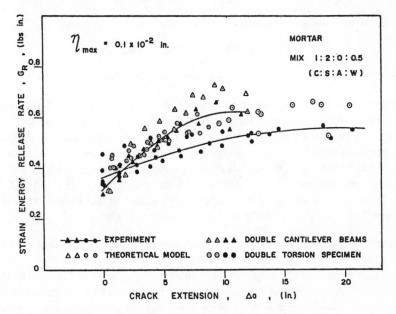

Figure 4.18.   Experimental and theoretical resistance curves.

the two types of specimens. This is perhaps because these curves were plotted with the observed crack extension Δa as an abscissa. However, to include the effect of the process zone, the R-curve should be plotted with the effective crack extension Δa_{eff} as the abscissa. This is done in Figure 4.19. Now a unique R-curve is obtained. This means that the plot of strain energy release rate (modified to include the nonlinearity as explained above) vs. the effective crack extension $\Delta(a + l_p)$ may be a unique material property. To examine this, some available data from notched-beam tests were analyzed using the theoretical model.

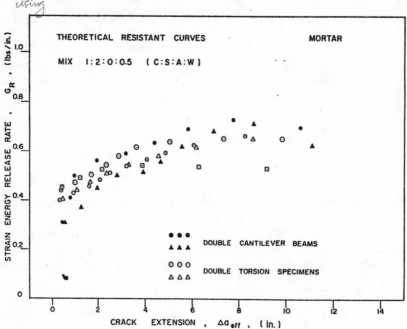

Figure 4.19.   A plot of G_R vs. Δa_{eff} for double cantilever and double torsion specimens.

Comparison with load-CMD curves for beams. Theoretical predicted values of load vs. crack mouth displacement of beams are compared with experimental data in Figure 4.20. For a given load and crack length, crack opening displacement was calculated based on global compliance and equations developed by Okamura [31]. These equations are valid only for a/W less than 0.6 whereas the observed crack length, more often than not, was greater than this value. A more accurate prediction of COD for notched beams is the subject of the current investigation at Northwestern University. The theoretical values of process zone length are shown in Figure 4.17.

Comparison of data on beams tested by other investigators. It is possible to calculate the load at crack propagation for notched beam specimens using the concept of R-curves mentioned earlier. Assuming that the R-curve reported in Figure 4.19 for the double cantilever and double torsion specimens is also valid for various beam specimens one can predict when a given crack length a, will propagate. This is shown in Figure 4.21 for a set of beams tested by Walsh [32]. Similar comparison was also

130

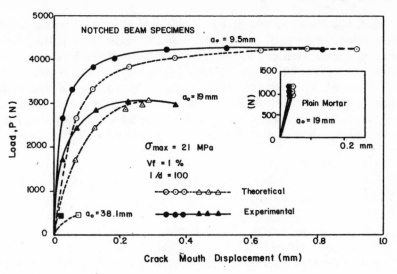

Figure 4.20.   Theoretical and experimental load-CMD curves.

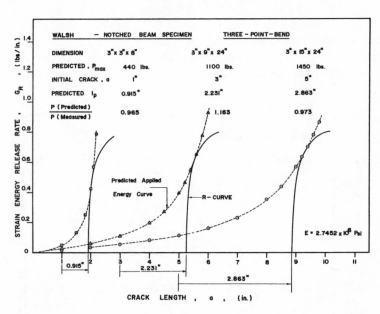

Figure 4.21.   Theoretical and experimental fracture loads for notched beams.

made with the results of several investigators [33–37]. The solid lines in the figure represent R-curves (taken from Figure 4.19). These curves are a plot of G_R vs. Δa_{eff}. If it is assumed that the process zone does not change with crack growth, then they can be plotted with respect to Δa by simply shifting the curve by an amount equal to the corresponding value of l_p. The values of l_p were calculated theoretically for these beams (Figure 4.17). The dashed line represents change in applied energy, G_I

131

with respect to a for a constant value of P. The applied strain energy release rate was calculated using the formula developed by Brown and Srawley [38]. The value of the modulus of elasticity was calculated either from the reported E values or from the values of f'_c $(E = 33W^{1.5} \sqrt{f'_c}; W = 145\,\text{lb/ft}^3)$ provided by the investigators [32–36]. If both E and f'_c were not given [37], a value of 3×10^6 psi was assumed for the modulus of elasticity. When the dashed curve is tangent to the R-curve, the value of load at unstable crack propagation is obtained. The theoretical values are compared with the measured maximum loads in Figure 4.21. A reasonably close comparison is observed.

4.7 Effects of compositions

R-curves for mortar and concrete DT specimens are shown in Figure 4.22. It can be seen that the higher the maximum size or volume fraction of aggregates, the higher is the value of fracture toughness as measured by R-curves. It should be noted that the conventional compressive strength of concrete specimens was lower than that of mortar specimens, but the values of fracture toughness as measured by the proposed R-curve method are higher for concrete specimens. On the other hand, the reported values of fracture toughness, calculated using LEFM, generally show a direct relationship with the corresponding strength values.

Theoretical plots of strain energy release rate vs. crack extension (R-curves) for the double cantilever (fiber reinforced) beam specimens are shown in Figure 4.23. The

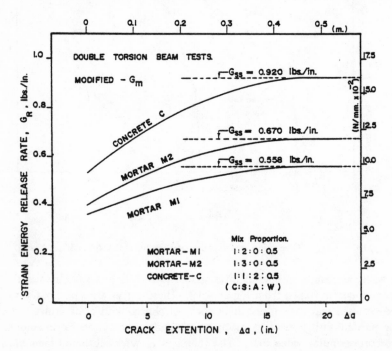

Figure 4.22. R-curves for mortar and concrete.

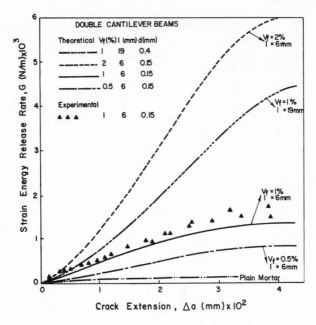

Figure 4.23.   R-curves for fiber reinforced concrete.

experimental R-curves were available for only one set of specimens; only these specimens were unloaded and reloaded after each observation of crack extension. The comparison between theory and experimental results is judged satisfactory.

Note that if the asymptotic value of the R-curve is considered a material parameter, then that value can be a useful quantity in identifying the benefits of fiber addition. For example, the steady state asymptotic value of specimens reinforced with 37.5 mm fibers is approximately 40 times that for plain, unreinforced matrix (Figure 4.23). This relative improvement in fracture energy is comparable to the reported value of 'toughness index' (relative values of the area under the load-deflection curves in flexure) [39].

Note that the shape of the R-curves is likely to depend on the specimen geometry. It should be possible to predict the fracture resistance curves using the method proposed here for various geometry of the specimens for different matrix compositions and for different properties of fibers.

Acknowledgment

The research reported in this chapter is being supported by grants from U.S. Air Force Office of Scientific Research, National Science Foundation and U.S. Army Research Office to Northwestern University.

References

[1]   Velazco, G., Visalvanich, K., and Shah, S.P., Fracture Behavior and Analysis of Fiber Reinforced Concrete Beams, *Cement and Concrete Research*, 10, pp. 41–51 (1980).

133

[2]   Shah, S.P. and Chandra, J., Critical Stress, Volume Change, and Microcracking of Concrete, *Journal of the American Concrete Institute*, 65, pp. 770–781 (1968).

[3]   Shah, S.P., and Slate, F.O., Internal Microcracking, Mortar-aggregate Bond and the Stress-Strain Curve of Concrete, *The Structure of Concrete* (International Conference, London, 1965); Cement and Concrete Association, London, pp. 82–92, (1968).

[4]   Shah, S.P., and Winter, G., Inelastic Behavior and Fracture of Concrete, Couses, *Mechanism and Control of Cracking in Concrete*, SP-20, American Concrete Institute, pp. 5–28 (1968).

[5]   Shah, S.P., and McGarry, F.J., Griffith Fracture Criterion and Concrete, *Journal of the Engineering Mechanics Division*, ASCE, EM6, pp. 1663–1676 (1971).

[6]   Wecharatana, M., and Shah, S.P., Double Torsion Tests for Studying Slow Growth of Portland Cement Mortar, *Cement and Concrete Research* 10, pp. 833–844 (1980).

[7]   Wecharatana, M., and Shah, S.P., Slow Crack Growth in Cement Composites, *Journal of Structural Division*, ASCE, June 1982.

[8]   Wecharatana, M., and Shah, S.P., Prediction of Nonlinear Fracture Process Zone in Concrete, *Journal of the Engineering Mechanics Division*, ASCE, to be published.

[9]   Wecharatana, M., and Shah, S.P., A Model for Predicting Fracture Resistance of Fiber Reinforced Concrete, *Cement and Concrete Research*, (submitted).

[10]   Fracture Toughness Evaluation by *R*-curve Methods, ASTM-STP 692, ASTM, Philadelphia 1973.

[11]   Mai, Y.W., Strength and Fracture Properties of Asbestos-Cement Mortar Composites, *Journal of Material Science*, 14, pp. 2091–2102 (1979).

[12]   Lenian, J.C., and Bunsell, A.R., The Resistance to Crack Growth of Asbestos Cement, *Journal of Material Science*, 14, pp. 321–332 (1979).

[13]   Andonian, R., Mai, Y.W., and Cotterell, B., Strength and Fracture Properties of Cellulose Fiber Reinforced Cement Composites, *Int. J. Cement Composites*, 1, pp. 151–158 (1979).

[14]   Brown, J.H., Measuring the Fracture Toughness of Cement Paste and Mortar, *Magazine of Concrete Research*, 24, pp. 185–196 (1972).

[15]   Irwin, G.R. and Nies, J.A., A Critical Energy Rate Analysis of Fracture Strength, *Welding Research Supplement*, 1954, pp. 195–198.

[16]   Hillemeier, B., and Hilsdorf, H.M., Fracture Mechanics Studies on Concrete Compounds, *Cement and Concrete Research*, 24, pp. 185–196 (1977).

[17]   Dugdale, D.S., Yielding of Steel Sheets Containing Slits, *Journal of Mechanics & Physics of Solids*, 8, pp. 100–104 (1960).

[18]   Williams, J.G., Visco-Elastic and Thermal Effect on Crack Growth in PMMA, *Inter. Journal of Fracture*, 8, pp. 393–401 (1972).

[19]   Barenblatt, G.J., The Mathematical Theory of Equilibrium Crack in the Brittle Fracture, *Advance in Applied Mechanics*, 7, pp. 55–125 (1962).

[20]   Foote, R.M.L., Cotterell, B., and Mai, Y.W., Crack Growth Resistance for a Cement Composite, *Advances in Cement-Matrix Composites*, Proceedings Symposium, Material Research Society, Annual Meeting, Boston, Massachusetts, pp. 135–144 (1980).

[21]   Petersson, P.E., Fracture Mechanical Calculations and Tests for Fiber Reinforced Cementitious Materials, *Advances in Cement-Matrix Composites*, Proceedings, Symposium L, Materials Research Society, Annual Meeting, Boston, Massachusetts, pp. 95–106 (1980).

[22]   Shah, S.P., Stroeven, P., Dolhuisen, D., and Stakelengerg, P. von, Complete Stress-Strain Curves for Steel Fiber Reinforced Concrete in Uniaxial Tension and Compression, Proceedings, International Symposium, RILEM–ACI=ASTM, Sheffield, pp. 399–408 (1978).

[23]   Naaman, A.E., Moavenzadeh, F., and McGarry, F.J., Probabilistic Analysis of Fiber Reinforced Concrete, Proceedings of ASCE, *Journal of Eng. Mech. Div.*, 100, pp. 397 (1974).

[24]   Visalvanich, K., Ph.D. *Dissertation*, Dept. of Material Engineering, University of Illinois, Chicago, 1982.

[25]   Swamy, R.N., Manget, P.J., and Roa, C.V.S.V., The Mechanics of Fiber Reinforcement of Cement Matrices, in *Fiber Reinforced Concrete*, ACI Publication SP-49, pp. 1–28 (1974).

[26]   William, D.P., and Evans, A.G., A simple Method for Studying Slow Crack Growth, *Journal of Testing and Evaluation*, 1, pp. 264–276 (1973).

[27]   Fuller, E.R., An Evaluation of Double Torsion Testing, *Analysis in Fracture Mechanics Applied to Brittle Materials*, ASTM – STP 678, pp. 3–18 (1979) (S.W. Frieman, Editor).

[28] Wiederhorn, S.M., Shorb, A.M., and Moses, R.L., Critical Analysis of the Theory of the Double Cantilever Method of Measuring Fracture – Surface Energies, *J. of Applied Physics*, 39, pp. 1562–1572 (1968).

[29] Lott, J.L., Kesler, C.E., and Naus, D.J., Fracture Mechanics – Its Applicability to Concrete, *Mechanical Behavior of Materials*, Vol. IV, Society of Materials Science, Japan. pp. 113–124 (1972).

[30] Schinker, M.G., and Doll, W., Interference Optical Measurements of Large Deformations at the Tip of a Running Crack in a Glassy Thermoplastic, *Mechanical Properties of Materials at High Rates of Strain*, Ed., J. Harding, Inst. Phys. Conf. Ser., No. 47, Chapter 2, pp. 224–232 (1979).

[31] Okamura, H., Watanabe, K., and Takano, T., Deformation and Strength of Crack Member Under Bending Moment and Axial Force, *Engineering Fracture Mechanics*, 7, pp. 531–539 (1975).

[32] Walsh, P.F., Fracture of Plain Concrete, *Indian Concrete Journal*, 46, pp. 469–470, 476 (1972).

[33] Mindess, S., Lawrence, F.V. and Kesler, C.E., The J-Integral as a Fracture Criterion for Fiber Reinforced Concrete, *Cement and Concrete Research*, 7, pp. 731–742 (1977).

[34] Kaplan, M.F., Crack Propagation and the Fracture of Concrete, Proceedings, *ACI Journal*, 58, pp. 591–610 (1961).

[35] Naus, D.J., and Lott, J.L., Fracture Toughness of Portland Cement Concretes, *ACI Journal*, pp. 481–489 (1969).

[36] Swartz, S.E., Hu, K.K., Fiartach, M., and Huang, C.J., Stress Intensity Factor for Plain Concrete in Bending – Prenotched versus Precracked Beams, (to be published).

[37] Huang, C.J., Swartz, S.E., and Hu, K.K., On the Experimental and Numerical Analysis of Fracture Toughness of Plain Concrete Beams ASTM *Symposium on Fracture Mechanics Methods for Ceramics, Rocks and Concrete*, Chicago, Illinois (1980).

[38] Brown, W.F., Jr., and Srawley, J.E., Plain Strain Crack Toughness Testing of High Strength Metallic Materials, ASTM-STP 410, ASTM, Philadelphia (1967).

[39] Henager, C.H., A Toughness Index for Fiber Concrete, RILEM *Symposium*, edited by R.N. Swamy, pp. 79–86 (1978).

5

*Microcracking in concrete**

5.1 Introduction

The effects of internal, microscopic defects or discontinuities on the macroscopic behavior of materials is an important study in itself. Although materials scientists have long worked on the effects of dislocations in metals, crystal defects in ceramics, and internal cracking in glass, it has only been relatively recently that attention has been given to the characterization of internal defects in concrete.

Concrete is a heterogenous, multiphase system which on a macroscopic scale is a mixture of cement paste, of fine aggregate in a range of sizes and shapes, of large aggregates in a range of gradation, and of a variety of types of void spaces. At the microscopic and submicroscopic levels the heterogeneity is pronounced, as the paste is observed to be a mixture of different types of crystalline structures at varying degrees of hydration collectively forming an amorphous gel. This mass contains voids due to the loss of water to both hydration of the cement particles and evaporation, as well as entrained and entrapped air voids. For the purpose of studying the structural behavior of concrete this complex mass can be simplified to a two-phase composite structure consisting of a coherent mortar phase and fine aggregate, bonded to the aggregate phase which is the coarse or large aggregate itself. To complete the model for this discussion, defects known as 'microcracks' are imposed upon the two-phase composite. Bond cracks occur at the mortar-aggregate interfaces, while mortar and aggregate cracks occur in their respective phases. The existence of these microcracks had earlier been suggested and assumed by several researchers [5, 13, 17, 23, 27, 35], but it was not until the topic was investigated thoroughly at Cornell Universtiy that such cracks were carefully observed, measured, and characterized [18–21, 46], in interior portions of the system.

Brittle nature of concrete (despite shape of the stress-strain curve). The typical stress-strain curve for concrete generally resembles that of many useful engineering materials

---

*The study of the character and mechanism of microcracking in concrete has been conducted by researchers in many institutions and in several countries. In lieu of presenting a comprehensive review of all such work, the authors have prepared a general summary of the work in the field with particular attention to the work performed at Cornell University from the early 1960's to the present.

with a linear region of apparently elastic behavior, followed by a curvature tending towards plastic behavior (Figure 5.1). Because of this classic shape of the stress-strain curve, under certain circumstances concrete has been classified as a 'ductile' material. Although some aspects of the macroscopic behavior of concrete members under load can be described as ductile, such behavior is, in the opinion of the authors, actually the result of a large number of microscopic brittle failures. Figure 5.2 shows a magnified portion of the stress-strain curve indicating the many discrete, irreversible brittle failures that accompany increasing stress and strain. The conventional stress-strain curve is actually an approximation or simplification of the true 'stepwise' stress-strain relationship.

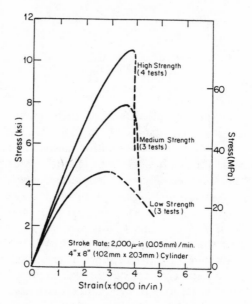

Figure 5.1.   Typical stress-strain curve of low-, medium-, and high-strength concretes. (From Carrasquillo et al., [11]).

Each of the steps shown in Figure 5.2 is manifested by the occurrence of a microcrack. The load paths in the specimen are therefore constantly changing and as the number of available load paths decreases, the intensity of load on the remaining paths rapidly increases. The resulting strain increases faster than the stress from the applied load and the stress-strain curve begin to curve to the right in imitation of plastic-ductile behavior [21, 41].

Tensile mode of failure of concrete. Although concrete is generally used under compression, the predominant failure mode is tension, or tension-shear. Even though the compressive strength of concrete is defined by the results of the standard cylinder compression test, the test itself produces a tension or tension-shear failure in the specimen in lieu of a true compressive crushing [51].

A failure mode of principal importance to concrete is the transverse tensile strain resulting from longitudinal compressive stress. In concrete under compression testing,

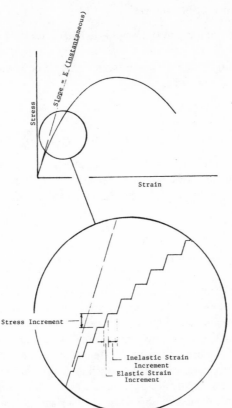

Figure 5.2.   Magnified portion of typical stress-strain curve of concrete showing both elastic and inelastic components of stress-strain behavior. Elastic response is assumed to act at the initial, or 'instantaneous' modulus of elasticity.

when no lateral restraint is present, the failure mode is tension splitting parallel to the direction of the load. As will be discussed further in the section on Multi-Axial Stress, the most reliable failure criterion for concrete may be a limiting tensile strain. Gilkey [16] suggested that lateral strain of compression test cylinders may be a more meaningful parameter than longitudinal strain.

It is evident, therefore, that the relatively low tensile strength of concrete directly or indirectly controls much of the behavior of the material. As will be seen, the existence of microcracking prior to the application of loads is largely responsible for the low tensile strength.

Early work on cracking of concrete. As early as 1929 Richart et al. [35] suspected the existence of internal cracking in concrete below maximum load. In particular, he observed an increase in the volume of compression test cylinders at a 'critical load' which corresponded to 77 to 85% of the maximum compressive load. He attributed this 'swelling' to microscopic internal splitting throughout the specimen. In 1952, Jones [23] determined that the velocity of sound through concrete changes when a specimen is loaded to 25 to 30% of its ultimate load. This change in sound velocity indicates a change in the internal structure of the specimen with increasing load.

Similar results were obtained by L'Hermite [27] and Rüsch [38] in 1954 and 1959 respectively.

In 1955, Blakey [6] postulated what seems now to be obvious, that the 'cracking of concrete may start in one of three ways:

(1) Fracture of cement paste
(2) Fracture of aggregate
(3) Failure of bond between the paste and the aggregate.'

Although all of these researchers had correctly suspected the occurrence of the internal microcracking, it was not until the early 1960's that the work at Cornell led to the direct observation and quantitative study of microcracking, as will be described.

5.2 Microcracking of concrete

Exterior vs. interior cracks. The matter of surface or exterior cracks vs. internal cracks must be put forward. The great majority of observations of cracks by magnification have been made on external surfaces only. Cracks and crack patterns on surfaces must be different, and sometimes much different, from internal cracking. After all, a free surface has an entirely different stress field than an interior region unless the entire body is stress free — which is never true in concrete. Therefore, loading will cause different stresses on a surface than in an interior, and thus different cracking. Further, any drying or shrinkage at a surface will cause different cracking at that surface as compared to the interior. Most experiments on concrete allow some surface drying. To make matters worse, often a thin layer of laitance will be present from contact with forms or from finishing, and often is not removed (as by grinding or cutting). Such a layer of laitance is especially sensitive to cracking from drying. Much of the work done on fracture mechanics of concrete involves use of a microscope to observe lengths of cracks on exterior surfaces of notched beams. Qualitatively, it is expected that the lengths of these exterior cracks will be longer than internal cracks above the notch, particularly if significant drying shrinkage has occurred. This matter should be given serious consideration.

Size limits of microcracks. Microcracks have not been generally defined as to size. The authors favor an upper limit of 0.1 mm for microcracks, to distinguish them from the larger macrocracks; this size represents an average limit of the unaided human eye, below which magnification is generally needed. A lower limit for microcracks would seem to be the smallest crack-like discontinuity that can be detected. This may be the tensile strain limit beyond which elementary building units of the brittle material (usually ions or dipoles for brittle materials) will not return to their original positions when load is released, thus resulting in a permanent separation between particles. This size is in the order of magnitude of ionic diameters. Thus the question of when and where a crack actually starts remains as an important gap in our knowledge. It is likely that we first detect cracks in concrete only after they are far larger than when they were first formed.

No-load cracks. As is suggested by the wide difference between compressive and tensile strengths of concrete, there exists in concrete a significant number of cracks or crack-like voids prior to the application of external loads. Most of these pre-load cracks are interfacial cracks between the mortar and coarse aggregate phases (a few of them may be fractures in the coarse aggregate, from crushing). Some of these interfacial cracks are caused by settlement of coarse aggregate during the placing process and an accompanying upward migration of mix water, known as 'bleeding.' As a result of this process, water accumulates on the undersides of aggregate particles forming a gap or at least a partial loss of bond between the mortar and aggregate (there is often a zone of higher water-cement ratio under the aggregate particles). Microscopic examinations of non-loaded specimens indicate that the bond cracks resulting from settlement or 'water gain' are only a portion of the bond cracks present, and that early volume changes can also produce cracks. This observation was pursued by Hsu et al. [18–21], and later by Slate and Matheus [44] as described below.

Volume-change bond cracks. Hsu [19] conceived a mathematical model consisting of rigid circular disks representing coarse aggregate surrounded by a medium representing the mortar. By the application of numerical techniques, he was able to estimate that tensile stresses at the mortar-aggregate interface of about 1800 psi can result from a shrinkage strain of 0.3% in the mortar. Laboratory testing by Hsu and Slate [20] on the tensile bond strength at the mortar-aggregate interface showed that the bond strength varied from 33 to 67% of the tensile strength of the mortar. For an average concrete mortar, this results in a tensile bond strength of interface of about 150 to 250 psi, clearly well below the strength levels required to resist cracking due to shrinkage of the mortar. This phase of the investigation was concluded by Slate and Matheus [44] verifying that mortar phase shrinkage strains of 0.3% and larger do occur. It may therefore be concluded that the occurrence of bond cracking at the mortar-aggregate interface is due not only to settlement and water gain, but also to volume changes during setting and hardening.

Short-term loading cracks. As previously described, the cracks formed prior to the application of external load are primarily bond cracks at the mortar-aggregate interface, with negligible cracking in either the mortar or aggregate phases. (Observation of aggregate cracking prior to the application of load would be indicative of unsound or previously fractured aggregate, while observation of significant amounts of pre-load mortar cracking might be indicative of freeze-thaw damage, sulphate attack, alkali aggregate reaction, etc.) At applied loads of up to about 30% of the ultimate load, the increase in bond cracking is negligible. This stress range corresponds to the linear or 'elastic' portion of the stress-strain curve.

As the load is increased the existing bond cracks increase in length and width and new bond cracks are formed, occurring first around the larger aggregate particles. (This observation corresponds with Alexander and Wardlaw's [1] conclusion that mortar to aggregate bond strength varies inversely with aggregate size.)

Using Acoustic Emission techniques, Jones [23] and Rüsch [38] independently reported increased internal cracking in the load range of 25 to 50% of ultimate. As shown by Hsu et al. [21], it is not until the load reaches 70 to 90% of ultimate, how-

141

ever, that cracks through the mortar increase noticeably. These mortar cracks bridge between adjacent bond cracks, generally at the points of shortest distance between aggregate particles. The mortar cracks and bond cracks therefore form a continuous crack pattern extending along and between several aggregate particles.

The development of continuous crack patterns is indicative of widespread cracking throughout the specimen. This also corresponds to the rapid increase in cracking noises recorded by Rüsch at approximately 75% of ultimate load, and the volumeric expansion of compression specimens observed by Brandtzaeg at 75 to 80% of ultimate. As load is further increased, the network of continuous bond and mortar cracks grows, forcing the applied load to follow fewer and fewer load paths until the internal structure is virtually disintegrated and disruptive failure results. Even at very high strains and immediately prior to failure, there is very little cracking in the aggregate particles themselves in normal-strength concrete, although considerable cracking through the coarse aggregate does occur in high-strength concrete.

Long-term loading cracks. The development of extensively interconnected cracking and expansion of the specimen occurs at what is called the 'critical' load. This is generally at 70 to 90% of ultimate in a short-term test, although it has been variously reported at between 71 and 96% [17, 35, 38]. This point (or load range) also represents the long-term compressive strength of concrete.

Prior to reaching the critical load, the cracks are predominantly bond cracks which are stable and do not increase in size or number under a constant load. In the critical load range, mortar cracks begin to develop. Mortar cracks become stable only when they are stopped at each end by bond cracks or other crack arrestors. Until the mortar crack is stopped, it is unstable and will continue to grow under a constant load [12, 29]. Therefore, due to the instability of unarrested mortar cracks, the concrete specimen will continue to deform until failure at a sustained load high enough to initiate the unstable mortar cracking. The point at which the unstable form of mortar cracking begins is known as the 'discontinuity point' [29]. Concrete loaded up to or above the discontinuity point (critical load) will fail under sustained stress. In normal-strength concrete this point is generally 70 to 75% f_c'.

5.3 Methods of study of microcracking

The investigation of 'microcracking' ranges from a macroscopic study of the behavior of cracked specimens to a microscopic (literally) study of the cracks themselves. As previously noted, the presence of microcracks was predicted on the basis of macro-behavior [25, 35, 38] and verified by microscopic study [18–21].

Sonic testing. One of the principal means of quantifying the presence of internal microcracking in concrete is through the application of Sonic Testing. This testing is based upon measuring the apparent velocity of sound waves through an elastic continuum, and involving such other things as resonant frequency.

Sound waves passing through an uncracked concrete specimen will do so at a specific velocity, depending on the specimen geometry and the density of the concrete.

A crack in the specimen, however, represents a discontinuity across which a sound wave cannot pass efficiently. As a result, sound waves in cracked specimens must pass around microcracks thus traveling a further distance than in the uncracked distance (or travel more slowly through the air or other fluid in the crack). The cracked specimen travel time will therefore be greater than in the uncracked specimen. Whitehurst [52] has described the above principles in detail and further discusses applications, equipment, and interpretation of results.

Using the 'Pulse Velocity' methods outlined above, one can record the increase in sound wave travel time with increase in cracking as compressive load is increased on a concrete specimen. In addition to detecting the presence of cracking (or more properly, discontinuities in the elastic medium) sonic testing can be used to determine elastic material properties such as the modulus of elasticity and Poisson's ratio. When these values are determined by sonic testing they are generally referred to as the 'Dynamic' moduli to differentiate them from the traditional 'static' moduli. A thorough discussion of the techniques required is found in [52].

Another related test method is known as Acoustic Emission. In simplest terms, the investigator 'listens' to the test specimen to detect the occurrence of internal cracking. The sophistication of the test varies depending on the type and sensitivity of the acoustic emission amplification, collecting, and recording equipment. Both L'Hermite [27] and Jones [23] used a form of acoustic emission testing in their work in internal cracking.

As a final comment on sonic testing, Slate has informally suggested an Acoustic Absorption Spectrum test, in which a selected spectrum of sound waves at varying frequencies would be introduced to a concrete specimen, and the frequencies of the response separated and measured. Such a test would seek to identify not only the presence of microcracking but also the type or size of microcracks as indicated by particular absorbed or scattered frequencies.

Microscope and X-ray techniques. Starting about 1960, intensive study of cracks in concrete was begun at Cornell University. The first motivation was to attempt to explain the shape of the stress-strain curve in terms of the internal structure and in terms of changes in the internal structure with loading. It was felt necessary to observe the interior of the concrete, instead of the surface of specimens.

A decision was made to use direct observation of internal structures and cracks, and not to use the interesting but indirect methods involving sound energy and sound emission upon cracking, as developed and used by Jones [23] and by Rüsch [38]. This resulted in the necessity of using destructive tests, to expose the internal structure. Thus specimens were loaded or treated in the desired manner, then cut open for observation.

Methods studied and discarded. Several methods were studied at Cornell University, either briefly or extensively — all methods studied involved radiant energy. Among methods studied briefly and then discarded were the following:

(1) Cut and polish a surface, allow a hydrophilic tracer liquid to penetrate cracks and voids in the slightly dried surface, lightly grind the surface to remove the surface

143

film of the tracer liquid, then examine the surface for cracks and voids as shown by the tracer liquid which had penetrated them. Tracers tried were fluorescent aqueous systems, and mildly radioactive aqueous solutions; the former could be observed or photographed directly in a darkroom under ultraviolet radiation, and the latter could be recorded by placing photographic film directly on the surface with cracks containing radioactive materials that emitted short wave radiation. The fluorescence method was somewhat useful, but cumbersome, and not conducive to detailed prolonged study such as can be done readily with the optical microscope. The radioactive method did not work well.

(2) Cut a slice of perhaps one-half to one centimeter thickness, allow it to dry a little, place the bottom face in a container of an aqueous dye that did not reach to the top of the slice, allow capillary rise of the dye to bring it to the top surface through cracks, then observe or photograph the colored cracks. This did not work well because the internal surfaces of the cracks adsorbed most of the dyes during capillary rise, and because (as was later learned) many or most of the cracks may not be continuous for an appreciable distance, even through the thickness of the thin specimen.

(3) Cut a thin slice, partially dry it, allow capillary rise of a saturated water solution of a lead salt, then X-ray to see cracks and voids that were then relatively opaque. This did not work well.

After other minor trials, two major methods were developed, one involving use of transmitted X-rays and the other involving optical (microscopic) study of a surface in which a dye had penetrated cracks and voids open to the surface. Apparently the X-ray technique had not been used previously for study of concrete; it is described in detail by Slate and Olsefski [46] in their 1963 paper on 'X-rays for Study of Internal Structure and Microcracking of Concrete.' This paper also described the optical method involving dye absorption. Another paper in 1963 by Hsu et al. [21] describes the optical method in somewhat different terms and refers briefly to the X-ray method. Details of the microscope technique with dye are given below, and those of the X-ray technique follow.

Microscope technique with dye. The following description is adapted from Slate and Olsefski [46]:

After processing to cause cracking, or after simply curing, a diamond rotary saw was used to cut horizontal or vertical slices 0.150 to 3 in. thick. The thinner slices were used when both microscope and X-ray techniques were to be used on the same specimen, as checks.

The coolant for the saw was a mixture of equal parts by volume of de-odorized kerosene and light flushing oil. If the specimens are even slightly damp, the internal porous structure is hydrophilic and will not absorb the hydrophobic organic coolant; the coolant on the surface can easily be flushed away by flowing water. It is vital to remove the powder from the cutting process that has been packed into the cracks; this can be done by flushing out the powder (or paste) with a jet of water. Failure to do this can result in failure to detect cracks; in the authors' opinion, this has frequently happened in some work.

Direct microscopic observation of the cut surfaces of the slices was used. A saw-cut

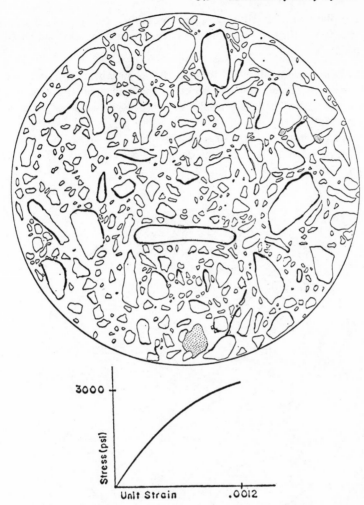

Figure 5.3. Sketch of a polished surface of concrete with cracks enhanced in ink by thick black lines, as observed through the microscope using the microscope-dye method; the section is horizontal and perpendicular to compressive load on a 4-inch diameter cylinder. The unit strain was 0.0012, as shown by the accompanying stress-strain curve. (From Hsu et al., [21]).

face of a specimen 0.150 to 3 in. thick was ground wet with abrasives (aluminum oxide or silicon carbide), starting with coarser abrasives and ending with No. 180 grit or smaller, on a plate glass surface, washed clean, then surface dried and painted with carmine drawing ink, which penetrated into voids and cracks. After the ink became surface dry on all the paste portions, the inked surface was ground wet with No. 180 or finer grit aluminum oxide or silicon carbide on a flat glass surface, until only a faint pink color could be seen by the naked eye. The drying time for the ink was about 15 min. Shorter drying time resulted in loss of most of the color into the water of the grinding medium; longer drying times resulted in formation of a tough surface film by the binder in the ink, and flaking and lifting of the colored material from larger cracks as the surface film was removed. Cracks and voids thus were dyed carmine, with great contrast to the rest of the surface of the specimen. This was the most successful dyeing technique tried.

A stereomicroscope was used, at 4X to 40X. Higher magnifications, and a monocular microscope, were employed only as checks on the technique; the stereomicroscope was much easier to use, and gave results substantially as good, or better for some purposes (e.g., depth).

Cracks were drawn on a sketch or a photograph of the surface being observed, as shown in Figures 5.3 and 5.4, and often compared with the X-ray plate of the same slice as a control.

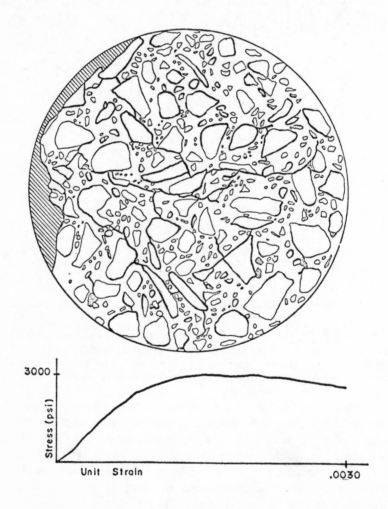

Figure 5.4.   Sketch of a polished surface of concrete with cracks enhanced in ink by thick black lines, as observed through the microscope using the microscope-dye method; the section is horizontal and perpendicular to compressive load on a 4-inch diameter cylinder. The unit strain was 0.0030, as shown by the accompanying stress-strain curve. (From Hsu et al., [21]).

X-ray technique. As far as the authors can determine, the first use of X-radiography to study cracks and other internal structural features of concrete (in contrast to X-ray diffraction and similar techniques for analytical study of composition) was that developed and used by Slate and Olsefski [46] and later used by their co-workers and associates Hsu et al. [21], Sturman et al. [48], Sturman et al. [49], Slate and Matheus [44], Shah and Slate [41], Meyers et al. [29], Buyukozturk et al. [8], Buyukozturk et al. [9], Liu et al. [28], Carino and Slate [10], Tasuji et al. [50], Carrasquillo et al. [11], and Ngab et al. [32]. Robinson [36] used X-rays at about the same time as Slate and Olsefski [46], but unfortunately used specimens that were too thick, (7 cm) resulting in radiographs that were of limited value, with impaired clarity and a confusing multitude of details. More recently, Isenberg [22], Bhargava [4], and Stroeven [47], among others, have done important work involving X-rays to study concrete. The technique is now well established.

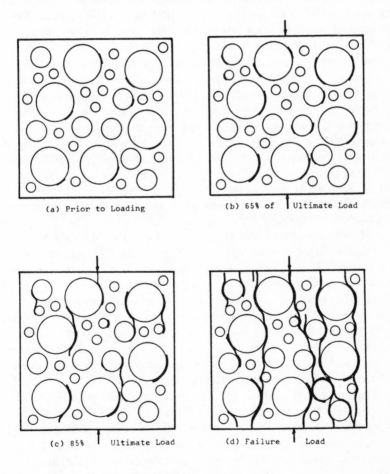

(a) Prior to Loading   (b) 65% of ⏐ Ultimate Load

(c) 85% ⏐ Ultimate Load   (d) Failure ⏐ Load

Figure 5.5.   Sketches of progressive cracking of a 1/2-inch thick plate of a model of concrete, made from X-ray images of the cracking, at various stages of uniaxial compressive loading, (From Liu, [28]).

147

Since the Cornell researchers felt strongly that meaningful observations and studies of cracks must be on interior cracks within the concrete rather than on cracks on an exterior surface, the idea of using X-rays was developed for observations at depth within concrete. It was correctly anticipated that the technique could be used either for thin specimens (up to about 2 cm thick, but best for specimens about 0.4 to 0.5 cm thick), or for slices of concrete carefully cut from the interior of specimens. Thus this technique met the specific requirements for observations that had been established, both for a destructive test involving cutting open concrete, and for a non-destructive test on thin intact specimens, as shown in Figure 5.5.

The following is adapted from Slate and Olsefski [46]:

After processing to cause cracking, or after simply curing, thin (plate) specimens were X-rayed directly. For other specimens, when it was desired to observe the interior portion of a mass of concrete (such as a concrete test cylinder or prism), a diamond rotary saw was used to cut horizontal or vertical slices 0.150 in. thick, plus or minus a maximum of 0.005 in. With care, the saw was capable of an accuracy of plus or minus 0.002 in. Specimens that were too thick showed so much detail as to cause confusion in interpretation of the X-ray plate, or caused loss of definition of very small cracks extending only a small part of the way through the specimen; specimens that were too thin sometimes had cracks introduced by the sawing process. Nonparallel faces lead to variation in darkness of different parts of the plate.

The coolant for the saw was a mixture of equal parts by volume of de-odorized kerosone kerosene and light flushing oil. If the specimens are even slightly damp, the internal porous structure is hydrophilic and will not adsorb the hydrophobic organic coolant; the coolant on the surface can easily be flushed away by flowing water. The specimen slices can be X-rayed wet or dry, as desired. Normally, they are kept wet, stored immediately in plastic bags with some water, and X-rayed while remaining in the bags (to prevent drying and shrinkage cracking).

An industrial X-ray unit was used, with a rating of 150 kilovolts (kv). The tube had a tungsten target, a focal length of 3 ft, an inherent filtration equal to approximately 0.5 mm Al, and a limit of 6 milliamperes at 150 kv. An exposure time of 2 min at 40 kv and 5 milliamperes was used. A fine-grained X-ray film was used, to make possible enlargements of the images. The film was developed as instructed by the manufacturer. However, developing time was increased or decreased to increase or decrease contrast, as desried for special observation. Enlargements of 3 to 10X were sometimes made from the X-ray plates, to aid in location of small cracks and other fine structure. The plates or enlargements were examined on a light box. A 4 to 10X hand lens was sometimes used to magnify a portion of the image for close study. An example of an X-ray image of concrete near failure is shown in Figure 5.6.

For X-ray study, a faint photo print (double reversal so cracks will be in black, not white) from the X-ray film is convenient for recording cracking, as illustrated by Figure 5.7. This print will, of course, look somewhat different from the surface or a photograph of the surface, since it shows details at depth, and may be a bit confusing at first.

Check for introduction of cracks by sawing, processing, and handling. The following description is from Slate and Olsefski [46]:

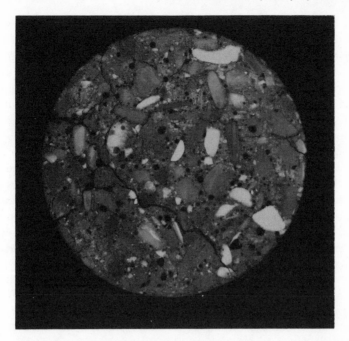

Figure 5.6.   X-ray image of a 0.15-inch thick slice of concrete cut from a 4-inch diameter cylinder and perpendicular to compressive loading. This image is a double reversal of the original X-ray film, so the cracks appear in black, as in the original. No enhancement of cracks or other structures was used. The concrete is near failure, and shows extensive cracking through the hydrated paste, as well as at paste-aggregate interfaces.

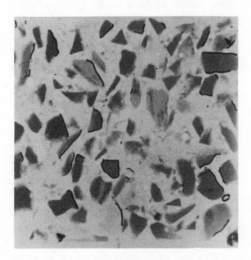

Figure 5.7.   Faint photographic print from an X-ray image of a slice of concrete cut from a 4-inch square prism tested under sustained load; the cracks are enhanced by thick black lines in ink. This image is a single reversal of the original X-ray film, so the cracks (before inking) appear in white and the more opaque aggregate particles appear in darker.

Slices 2 or 3 in. thick were cut from undisturbed cylinders fresh from the moist room or the curing tank. The sawed surface was stained, studied under the microscope, and a map made of all cracks present. A parallel cut was then made 0.150 in. from the first cut. The surface originally studied was restained, reground, and restudied with the microscope, and a new map of cracks made. No appreciable additional cracking was found on the second examination, after the thin slice has been sawed out of the larger mass. [This procedure required partial wetting and air-drying cycles – partial surface drying for application of ink, wetting by wet grinding, and partial surface drying (or inundation) for observation – which by themselves could cause cracking; the amount of drying was carefully kept to the absolute minimum needed by the techniques.] Further, any cracking caused by sawing must necessarily show directional effects from the pressure feed and direction of cut; careful examination showed the absence of any such directional pattern of cracking. Therefore, no appreciable additional cracking was introduced by sawing, processing, and handling. It should be emphasized that reasonable care must be taken in these steps, or cracking may be introduced.

General comments. Slices of concrete are routinely stored in plastic bags, until further processing, to prevent drying and shrinkage cracking. Extended storage should be avoided to prevent autogenous healing of cracks. However, for the dyeing technique, surface drying must occur, but appreciable drying at depth should be avoided (unless drying shrinkage is being studied).

For recording cracking, it was found helpful to photograph and make a faint print of the surface, then to draw the cracks on the print from study with the microscope. This resulted in sharp contrast for the lines representing cracks, and aided in their measurement.

Little change has been made in subsequent years to the procedures described above, because they continue to work well.

5.4 Effects of microcracks on the properties of concrete.

General. Microcracking may be said to be the fundamental failure mechanism of concrete. All mechanical failure of concrete is preceded and accompanied by the development of bond, mortar, and aggregate cracks as described earlier in this chapter. Although the discussion thus far has been confined to failure arising from the application of external loads, microcracking occurs also as the direct result of internally generated forces such as the hydraulic pressure resulting from freezing pore water in freeze-thaw damage, or the expansive effects of sulphate attack, alkali-aggregate reaction, or the corrosion of reinforcing steel. The tensile stresses developed as a result of these phenomena eventually cause the disintegration of the concrete through the mechanism of microcracking. Furthermore, the presence of microcracks undoubtedly increases the permeability of the concrete and thus encourages more rapid deterioration.

Since microcracking is such a fundamental process, the study of the relationship of various concrete properties to microcracking serves to unify many of the complicated cause-and-effect relationships in concrete materials science. In essence, since micro-cracking is the fundamental mechanical failure mechanism of concrete, any process that causes or increases the extent of microcracking will bring the concrete closer to failure. A concrete specimen which exhibits a large quantity of microcracks as a result of freeze-thaw damage will be able to accept less load than an undamaged specimen. Likewise, freeze-thaw damage causes a discernible increase in the volume of a speci-men, just as if it has been loaded to critical load.

It has been observed that insufficient curing can lead to early drying shrinkage bond cracks as described earlier [11]. The presence of such 'no load' cracks may have little effect on the concrete at low stress levels, but will reduce the number of available alternate load paths above the discontinuity point, and may contribute to lower ultimate strength. Therefore, not only does inadequate curing affect the degree of hydration of the cement gel, but the additional drying shrinkage cracking thus intro-duced may reduce strength as well.

Microcracking and the shape of the stress-strain curve. As previously described bond cracking as a result of applied load in a uniaxial compression test begins to appear at about 30% of f'_c. Up to this point the concrete behaves essentially in an elastic manner, and all of the work of deformation is stored in the concrete specimen as elastic strain energy. As the compressive stress is increased above 30% f'_c the incidence of bond cracking at the mortar-aggregate interface begins to increase the deformation (strain) at any given stress, and to transform stored elastic strain energy into fractured surface energy of the new crack faces. This results in a departure from linearity and a curving to the right of the stress-strain curve. This post-proportional limit behavior may be represented as a two step process consisting of: 1) an incremental linear elastic response to the applied load with no cracking of the concrete, accompanied by: 2) an incremental inelastic deformation at constant load accompanied by microcracking. (The formation of bond cracks is somewhat time-dependent, and it is noted that the stress-strain curve for concrete rapidly loaded tends more towards a linear elastic response with a modulus of elasticity approaching the slope of the curve up to 30% f'_c.) This two-step process gives rise to the 'stair-step' form of the magnified stress-strain curve shown in Figure 5.2 in the introduction.

Upon reaching approximately 70% f'_c, mortar cracking begins, which is seen in the stress-strain curve as the point at which curvature to the right becomes more pro-nounced. It has been noted [12] that mortar cracks are 'unstable,' which means that more stored energy is released by their formation than is required to do the work of fracture. The excess energy available may then continue to drive the cracks without additional load. Thus new cracks are formed at constant load, which is the observed material behavior once mortar cracking is initiated.

A 'failure mechanism' for concrete is described in [29]. During the early loading stages, bond cracking occurs when the unconfined bond strength is exceeded. Com-pression-shear load can still be carried by the mortar-aggregate composite, however, by friction at the interfaces inclined to the load. This load path is seriously affected by the mortar cracking which loosens the mortar and reduces the friction. After this occurs, load is shifted to other, more stable, uncracked mortar and aggregate load paths. As the cracking progresses, the intensity of load on the remaining load paths increases, which of course hastens cracking. Furthermore, mortar cracks will continue to run without increased load until they are arrested by bond cracks or other crack arrestors.

The release of stored strain energy by cracking of the mortar is thus spontaneous, once initiated. Therefore, once the discontinuity point has been reached and mortar cracks are initiated, more and more stored elastic strain energy is transformed to cracked surface energy without additional load. The stress-strain curve therefore begins to descend, incrementally storing less strain energy with each increment of strain. Eventually the system is so disintegrated that it cannot store any more energy and it fails.

Under sustained load above about 80% of f'_c, plain concrete is *unstable* and will fail; thus the ascending portion of the stress-strain curve above about 80% of f'_c, and the entire descending portion of the curve represent *unstable* behavior of concrete under sustained load (which represents almost all uses of concrete). Stability can be achieved by providing load transfer mechanisms, as by reinforcement.

The behavior of high-strength concrete can be explained in the same general terms. Since bond cracking is inhibited by the greater tensile capacity of the paste, strain energy is not transformed into cracking energy, and the stress-strain curve remains essentially linear to beyond 85% f'_c. A small amount of bond cracking precedes arrival at the discontinuity point, as evidence by the slight curvature in the stress-strain curve. At the discontinuity point, unstable mortar cracks are initiated which are not arrested by bond cracks but pass through the aggregate instead. This rapid transformation of energy is evidenced by the sharp curvature and sharp downturn of the stress-strain curve, ending in sudden failure.

Repeated load. The performance of concrete under low-cycle repeated load has been studied and is described in [39] and [42]. In this case, the investigators were interested in finding the 'shake down limit,' or the maximum repeatable load. During this study, it was found that the 'shake down limit' is essentially the critical load as defined previously. During the first load cycles to less than the critical load, the stress-strain behavior was as previously described with the accompanying bond cracks. Upon unloading and subsequent loadings to the same limit, the concrete behaved essentially elastically ~~elasticity~~ with some residual strain owing to the first cycle cracking.

After five load cycles the load was increased to failure. It was found that as long as the cyclic load was held under the critical load, the ultimate strength was essentially unaffected. When cyclic loads were applied in excess of the critical load, such that mortar cracking developed, failure ensued in a very few cycles. It may be concluded, therefore, that repeated loads can be tolerated provided that they are of such a magnitude so as to produce bond cracking only. Further research with a higher number of cycles is needed, however. Neville [31] indicates that the type of behavior described herein, i.e. elastic response for all subsequent loadings, may be valid for only a certain number of cycles.

Creep. Ngab et al [32, 33] discuss the relationship between microcracking and the creep of concrete. Although changes in the distribution of moisture within a specimen are the primary factors relating to creep, it has been observed that long term creep is accompanied by (and caused partially by) the formation of microcracks.

Flexural stresses. In order to investigate the relationship between flexural stresses and microcracking, Sturman, Shah, and Winter [48] studied cylinders loaded in eccentric compression. It was determined that the strain gradient resulting from the eccentric load reduced the incidence of microcracking, with a pronounced reduction in mortar cracking. The reduction in cracking was not due to reduced strains in the eccentric specimens. When regions of equal strain were compared between eccentric and concentric specimens, the eccentric load case resulted in only about one-half of the cracking observed in the concentrically loaded specimen.

The retarding effect of the strain gradient on the mortar cracking was significant enough to cause a much slower curving in the stress-strain curve than obtained in concentric loading once mortar cracking had been initiated. Prior to this point, which corresponded to a strain of approximately 0.0017, the two stress-strain curves were practically identical. The result of the retardation of the mortar cracking was a 20% increase in the ultimate strength (short term) with a 50% increase in the strain at failure.

Although these results appear extremely favorable for application to flexural members, the reader is cautioned that the strain gradients imposed by eccentric load were generally in excess of those present in flexural members such as beams. The benefits of the strain gradient would therefore be expected to be significantly less in the case of real beams. It is interesting to note, however, that the work of Sturman, Shah, and Winter demonstrated a significant difference in the stress-strain curve of concrete in flexure compared to concentric compression [48]. It is also noted that if the presence of a strain gradient increased instead of retarded micro-cracking, the traditional compression test would be an unsatisfactory indicator of strength for most concrete structures.

Tensile strength. It is evident that since microcracking is in itself a tensile failure mechanism, it should have a significant effect on the tensile strength of concrete. As stated previously, however, unless one is speaking in terms of the crushing of confined concrete as in a bearing failure, the failure of the concrete is always in tension or tension-shear.

An investigation of the stress-strain curve concrete in uniaxial tension was undertaken by Al-Kubaisy and Young [2, 34]. The ascending branch of the curve had two distinct linear segments, the first one extending from the origin to around 30% of ultimate tensile strength at a strain of 0.000016 to 0.000030. At this 'discontinuity point,' pulse velocity testing indicated a sudden increase in bond cracking. The second linear segment continued to a load of about 75% of ultimate tensile strength where a sudden increase in mortar cracking occurred. Failure then followed acompanied by a sharp bending of the stress-strain curve. It would appear from this testing that a concrete specimen in tension will behave elastically up to a 'bond-cracking discontinuity point' at which most of the bond cracking suddenly occurs. The specimen, now with significant loss of aggregate-mortar bond, continues to carry load (at a lower tensile modulus of elasticity) apparently depending on the stiffness of the mortar without the help of the cracked mortar-aggregate interfaces. At the 'mortar-cracking discontinuity point' the mortar cracks begin to form, initiating failure of the specimen. It is interesting that the second discontinuity point bears approximately the same proportionate relation to the ultimate tensile strength as does the discontinuity point in compression, i.e. about 75% of ultimate.

In all of the compression testing described thus far, the use of gravel (rounded aggregate) vs. crushed stone has appeared to make only a small difference. The tensile strength, on the other hand, is significantly increased by the use of crushed stone [18]. In the work by Hsu and Slate [20] on the tensile bond strength, it was determined that increased surface roughness increased the bond strength and that limestone reacted chemically to increased bond strength. Furthermore, it would be expected that crushed stone would provide better mechanical interlock and friction capacity for the second linear segment of Al-Kubiasy's stress-strain curve.

It has also been noted that the tensile strength of concrete is more sensitive to curing conditions than is the compressive strength. This is likely due to bond cracks formed as a result of drying shrinkage.

Multi-axial stress states. Concrete fails in tension or tension-shear unless the configuration of the loading is such to prevent the development of tensile strains beyond a small and critical limit. In the case of the typical uniaxial compression test, transverse tensile strains are developed due to the Poisson expansion of the test specimen. (As previously described, this is demonstrated by the vertical splitting of the specimens when the test is arranged so as to prevent transverse restraint at the ends of the specimen [3, 10]). A study of the behavior of concrete when the Poisson expansion is restrained requires the application of a transverse loading, i.e. multi-axial stress.

Studies of multi-axial stress have been reported in [8, 10, 26, 28, 37] which included biaxial tension and compression.

In biaxial compression, the restraint against Poisson expansion eliminates tensile cracking and therefore results in higher ultimate strengths than in the uniaxial case. The strength increase for equal compression on each axis is on the order of 16% when loading heads are used which partially restrain expansion in the third direction. When low friction loading contacts at the heads are used the strength increase is much smaller [26]. As would be expected from the uniaxial case, the biaxial failure mode is again tensile splitting, and is in a direction parallel to the free or unrestrained surface.

In the case of biaxial compression tension, the application of a transverse tensile stress reduces the ultimate compressive strength. This can be understood as the superposition of direct tensile stress with its resultant cracking, added to the tensile cracking induced by the compression. Failure is initiated by the development of a single microcrack which is then opened by the applied tensile stress. For the case of compression loading with transverse tension, the discontinuity point is the point at which the first major microcrack appears [10].

In the case of biaxial tension, the concrete performs essentially as in uniaxial tension [26]. Under biaxial tension the first major crack, presumably a bond crack, represents the discontinuity point and simultaneously marks the beginning of essentially instantaneous failure [10]. The application of tensile stress eliminates the friction forces which maintain post-bond crack integrity in uniaxial and biaxial compression specimens.

5.5   Cracking of higher-strength vs. lower-strength concrete

Investigations of microcracking in 'High-Strength' concrete, i.e. $f_c' > 9000$ psi, were carried out in detail and are reported in [11, 12, 32, 33]. In these investigations, a typical high-strength mix had a cement content of 10.5 bags per cubic yard, a water-cement ratio of 0.32. It is important to realize that such a concrete has a much larger relative volume of paste, and a higher-strength paste than a more conventional, or normal-strength concrete. The result of these increases in paste volume and strength is increase of strength and closer elastic compatibility between the coarse aggregate and mortar phases. The resulting concrete is therefore more homogeneous in nature.

The increase in paste strength resulting from the low water-cement ratio greatly increases the tensile capacity of the paste and also increases the bond strength between the mortar and aggregate phases. This leads to a reduced incidence of bond cracking with and without load.

Compared to normal-strength concrete, there is considerably less cracking at equal stresses relative to (as a percentage of) ultimate in high-strength concrete. This is due the decreased tendency for bond cracking as described. As a result, high-strength concrete behaves in a more elastic manner than normal-strength concrete, with a steeper, more linear ascending branch of the stress-strain curve. Appreciable divergence from linearity does not occur until approximately 85 to 90% of f_c', and then it is relatively small. Initiation of significant mortar cracking begins at stresses above about 85% of f_c'. At this point however, the behavioral homogeneity of the material causes mortar cracks to grow and to pass through coarse aggregate particles instead of stopping at bond cracks. In essence, the strength and modulus of elasticity of the mortar phase is comparable to those of the coarse aggregate. Since the bond cracks do not serve as crack arrestors, the growth of the combined mortar-aggregate cracks is unstable and rapid. The failure of high-strength concrete therefore occurs suddenly, as shown by the sharp curvature of the stress-strain curve at maximum load.

This sudden failure only slightly above the discontinuity point of approximately 85 to 90% of f_c' is indicative of a brittle material and at first reading seems to be an undesirable characteristic of high-strength concrete. Although the failure of normal-strength concrete can be characterized as 'ductile — preceded by a slow flattening of the stress-strain curve — such a failure occurs under sustained load as low as 75% of ultimate. High-strength concrete, on the other hand, can sustain a load of about 85% of ultimate prior to failure. If both materials were loaded to a service load of, say, 50% f_c', there would be a greater margin of reserve strength in the high-strength concrete. Alternatively, if the load on both materials was increased to ultimate, the normal-strength concrete would give earlier warning of impending failure by showing greater strain and deflection before the load reached ultimate, and would absorb relatively more energy in doing so.

5.6   Summary

In the opinion of the authors, it is now well-establihed that concrete, which is a brittle material, fails in tension or tension-shear no matter how it is loaded (in tension, in compression, or in shear), as long as tensile strains are not prevented from exceeding tensile strain limits for fracture – such as some cases of confinement or multiaxial compression – and even then failure on a micro scale is still probably tensile. The old, and hopefully discarded, idea that concrete fails in shear in a compression test is wrong because the diagonal pattern of failure is caused by restraint by the loading heads with a complex loading pattern rather than simple uniaxial compression, and tensile splitting occurs parallel to the external compression load and within regions where tensile strains are not sufficiently restrained.

Thus tensile and tensile-shear cracking control the initiation, progress and culmination of loading failure in concrete in short-term loading, and control a large part of the process of creep and failure of concrete in long-term loading. Details of the cracking process are now fairly well-understood for many or most methods of loading of concrete, down to a magnification of about 100X; work is now being done on greater magnifications [15].

The distinction between external and internal cracks is important. External cracks may not represent the true conditions in the interior of a mass of concrete, and conclusions based on external cracks must be qualified – in some cases they may be misleading or even wrong.

Methods of study that are indirect – not direct observation of cracks themselves, but observation or effects or changes related to cracking – such as sonic studies and volume change studies, continue to be useful. Further development of sound and energy methods to locate cracks and to estimate the degree of cracking may be particularly fruitful.

The use of the microscope, especially on surfaces cut from the interior of a processed specimen, continues to be a technique of major importance. The major disadvantage is that such tests are destructive and require a new and different specimen for each variable or change being studied. Progressive observation of cracks on a free surface, as a test is continued, remains an interesting technique of limited value, with the great advantage that the test is non-destructive and several observations can be made on the same specimen. The use of higher magnifications is proceeding and is of considerable interest – a drawback is changes in the system caused by the technique itself, such as an environment of low pressure used in electron microscopy.

The use of X-rays has become a major tool to study cracks in concrete, both as a destructive test to study thin specimens cut from the interior of a processed larger specimen and as a non-destructive test to follow cracking in thin plates as loading progresses. Stereo X-radiography promises to be an important specialized technique of considerable importance and limited application.

Significant differences occur between microcracking behavior of normal-strength and high-strength concretes (below and above about 7500 to 9000 psi). High-strength concrete cracks significantly less under all conditions. The differences in cracking have been related to the differences in properties, and are generally responsible for those differences.

In general, studies relating changes in internal structure of materials, such as microcracking, with changes and differences in properties of those materials, have been highly fruitful and promise to continue to be highly fruitful.

References

[1] Alexander, K.M. and Wardlaw, J., Dependence of cement aggregate bond strength on size of aggregate, *Nature*, 187, 4733, pp. 230–231 (July 16, 1960).

[2] Al-Kubaisy, M.A. and Young, A.G., Failure of concrete under sustained tension, *Magazine of Concrete Research*, 27, 92, pp. 171–178 (Sept. 1975).

[3] Avram, C. et al., *Concrete strength and strains*, Elsevier Scientific Publishing Co., Amsterdam, Holland (1981).

[4] Bhargava, J., Nuclear and radiographic methods for the study of concrete, *Acta Polytechnica Scandinavica*, Civil Engineering and Building Construction Series 60, The Royal Swedish Acad. of Engr. Sciences, Stockholm, 103 pp. (1969).

[5] O'Berg, O.Y., *K Voprosu O. Prochnosti I Plastichnosti Beonta (On the problem of strength and plasticity of concrete)*, Dokladi Akademii Nauk USSR (Moscow) 70, 4, 617 pp. (1950).

[6] Blakey, F.A., Some considerations of the cracking or fracture of concrete, *Civil Engineering and Public Works Review*, 50, 586 (April 1955).

[7] Brooks, A.E. and Newman, K., The structure of concrete, *Proceedings* of an International Conference, London (Sept. 1975). Organized by the Concrete Materials Research Group, Imperial College of Science and Technology, Univeristy of London, in conjunction with the Cement and Concrete Association.

[8] Buyukozturk, O., Nilson, A.H. and Slate, F.O., Stress-strain response and fracture of a concrete model in biaxial loading, *Journal of the American Concrete Institute, Proceedings*, 68, 8, pp. 590–599 (Aug. 1971).

[9] Buyukozturk, O., Nilson, A.H. and Slate, F.O., Deformation and fracture of a particulate composite, *Journal of Engineering Mechanics Division, Proceedings of American Society of Civil Engineers*, 98, pp. 581–593 (1972).

[10] Carino, N.J. and Slate, F.O., Limiting tensile strain criterion for failure of concrete, *Journal of the American Concrete Institute, Proceedings*, 73, 3, pp. 160–165 (March 1976).

[11] Carrasquillo, R.L., Nilson, A.H. and Slate, F.O., Properties of high strength concrete subject to short term loads, *Journal of the American Concrete Institute, Proceedings*, 78, 3, pp. 171–178 (May–June 1981).

[12] Carrasquillo, R.L., Slate, F.O. and Nilson, A.H. Microcracking and behavior of high strength concrete, *Journal of the American Concrete Institute, Proceedings*, 78, 3, pp. 179–186 (May–June 1981).

[13] Czernin, W., *Chemistry and physics of cement for civil engineers*, Chemical Publishing Co., New York (1962).

[14] Darwin, D. and Steel, F.O., Effect of paste-aggregate bond strength on behavior of concrete, *Journal of Materials*, JMLSA, 5, 1, pp. 86–98 (March 1970).

[15] Diamond, S. and Mindess, S., Scanning electron microscope observations of cracking in portland cement paste, *Proceedings Seventh International Symposium on the Chemistry of Cement*, 3, VI-114-119, Paris (1980).

[16] Gilkey, H.J., Water cement ratio versus strength – another look, *Journal of the American Concrete Institute, Proceedings*, 57, 10, pp. 1287–1312 (April 1961).

[17] Hognestad, E., Hanson, N.W. and McHenry, D., Concrete stress distribution in ultimate strength design, *Journal of the American Concrete Institute, Proceedings*, 52, 4, pp. 455–479 (Dec. 1955).

[18] Hsu, T.T.C., *Microcracks between coarse aggregate and paste-mortar in concrete*, Ph.D. thesis, Cornell University, Ithaca, New York (June 1962).

[19] Hsu, T.T.C., Mathematical analysis of shrinkage stresses in a model of hardened concrete, *Journal of the American Concrete Institute, Proceedings*, 60, 3, pp. 371–390 (March 1963).

[20] Hsu, T.T.C. and Slate, F.O., Tensile bond strength between aggregate and cement paste or mortar, *Journal of the American Concrete Institute, Proceedings*, 60, 4, pp. 465–486 (April 1963).

[21] Hsu, T.T.C., Slate, F.O., Sturman, G.M. and Winter, G., Microcracking of plain concrete and the shape of the stress-strain curve, *Journal of the American Concrete Institute, Proceedings*, 60, 2, pp. 209–224 (Feb. 1963).

[22] Isenberg, J., A study of cracks in concrete by X-radiography, *RILEM Bull.* (New Series) 30, pp. 107–114 (March 1966).

[23] Jones, R., A method of studying the formation of cracks in a material subject to stresses, *British Journal of Applied Physics*, London, 3, 7, pp. 229–232 (July 1952).

[24] Kaplan, M.F., Crack propagation and the fracture of concrete, *Journal of the American Concrete Institute*, 58, 5, pp. 591–607 (Nov. 1961).

[25] Krishnaswamy, K.T., Microcracking of plain concrete under uniaxial compressive loading, *Indian Concrete Journal* (Bombay), 43, 4, pp. 143–145 (April 1969).

[26] Kupfer, H., Kilsdorf, H.K. and Rüsch, H., Behavior of concrete under biaxial stress, *Journal of the American Concrete Institute*, 66, 8, pp. 656–666 (Aug. 1969).

[27] L'Hermite, R., Present day ideas in concrete technology. Part 3: The failure of concrete, *RILEM Bulletin*, 18, pp. 27–38 (June 1954).

[28] Liu, T.C.Y., Nilson, A.H. and Slate, F.O., Stress-strain response and fracture of concrete in uniaxial and biaxial compression, *Journal of American Concrete Institute Proceedings*, 69, pp. 291–295 (1972).

[29] Meyers, B.L., Slate, F.O. and Winter, G., Relationship between time dependent deformation and microcracking of plain concrete, *Journal of the American Concrete Institute Proceedings*, 66, 1, pp. 60–68 (Jan. 1969).

[30] Mindess, S. and Young, J.F., *Concrete*, Prentice-Hall, Inc., Englewood Cliffs, New Jersey (1981).

[31] Neville, A.M., *Properties of concrete*, 3rd edition, Pitman Press, Great Britain (1981).

[32] Ngab, A.S., Nilson, A.H. and Slate, F.O., Shrinkage and creep of high strength concrete, *Journal of the American Concrete Institute, Proceedings*, 78, 4, pp. 255–261 (July–Aug. 1981).

[33] Ngab, A.S., Slate, F.O. and Nilson, A.H., Microcracking in high strength concrete, *Journal of the American Concrete Institute, Proceedings*, 78, 4, pp. 262–268 (July–Aug, 1981).

[34] Orchard, D.F., Concrete technology, 4th edition, Vol. 1 – *Properties of materials*, Applied Science Publishers, London, p. 297 (1979).

[35] Richart, F.E., Brandtzaeg, A. and Brown, L., A study of the failure of concrete under combined compressive stresses, *Bulletin No. 185*, University of Illinois Engineering Experiment Station, Urbana, 102 pp. (April 1929).

[36] Robinson, G.S., Methods of detecting the formation and propagation of microcracks in concrete, *Proceedings of the International Conference on the Structure of Concrete*, Imperial College, London (1965). *Cement and Concrete Association*, London, pp. 131–145 (1968).

[37] Rosenthal, I. and Glucklich, J., Strength of plain concrete under biaxial stress, *Journal of the American Concrete Institute*, 67, 11, pp. 903–913 (Nov. 1970).

[38] Rüsch, H., Physikalische fragen der betonprüfung, (Physical problems in the testing of concrete), *Zement-Kalk-Gips*, 12, 1, pp. 1–9 (1959).

[39] Shah, S.P., *Inelastic behavior and fracture of concrete*, Ph.D. thesis, Cornell University, Ithaca, New York (1965).

[40] Shah, S.P. and Chandra, S., Critical stress, volume change, and microcracking of concrete, *Journal of the American Concrete Institute, Proceedings*, 65, 9, pp. 770–781 (Sept. 1968).

[41] Shah, S.P. and Slate, F.O., Internal microcracking, mortar-aggregate bond and the stress-strain curve of concrete, *The Structure of Concrete*, Proceedings of an International Conference, London (A.E. Brooks and K. Newman, editors), pp. 82–92 (1965).

[42] Shah, S.P. and Winter, G., Inelastic behavior and fracture of concrete, *Journal of the American Concrete Institute, Proceedings*, 63, 9, pp. 925–930 (Sept. 1966).

[43] Slate, F.O., Introductory remarks and discussion, *The Structure of Concrete*, Proceedings of an International Conference, London (A.E. Brooks and K. Newman, editors) (Sept. 1975).

[44] Slate, F.O. and Matheus, R.F., Volume changes on setting and curing of cement paste and concrete from zero to seven days, *Journal of the American Concrete Institute, Proceedings*, 64, 1, pp. 34–39 (Jan. 1967).

[45] Slate, F.O. and Meyers, B.L., Deformations of plain concrete, *Proceedings, Fifth International Symposium on the Chemistry of Cement* (Tokyo 1968), Cement Association of Japan, Tokyo, 3, pp. 142–151 (1969).

[46] Slate, F.O. and Olsefski, S., X-ray for study of internal structure and microcracking of concrete, *Journal of the American Concrete Institute, Proceedings*, 60, 5, pp. 575–588 (May 1963).

[47] Stroeven, P., Morphometry of fiber reinforced cementitious materials. Part I: Efficiency and spacing in idealized structures, *Matériaux et Constructions*, 11, 61, pp. 31–38 (1978); Part II: Inhomogeneity, segregation and anisometry of partially oriented fiber structures, *Ibid*, 12, 67, pp. 9–20 (1979).

[48] Sturman, G.M., Shah, S.P. and Winter, G., Effects of flexural strain gradients in microcracking and stress-strain behavior of concrete *Journal of the American Concrete Institute, Proceedings*, 62, 7, pp. 805–822 (July 1965).

[49] Sturman, G.M., Shah, S.P. and Winter, G., Microcracking and inelastic behavior of concrete, *Proceedings of the International Symposium on Flexural Mechanics of Reinforced Concrete*, Miami, Florida (1964); *American Concrete Institute Special Publication*, 12, pp. 473–499 (1965).

[50] Tasuji, M.E., Slate, F.O. and Nilson, A.H., Stress-strain response and fracture of concrete in biaxial loading, *Journal of the American Concrete Institute, Proceedings*, 75, pp. 306–312 (1978).

[51] Troxell, G.E., Davis, H.E. and Kelly, J.W., *Composition and properties of concrete*, 2nd edition, McGraw-Hill, p. 228 (1968).

[52] Whitehurst, E.A., Evaluation of concrete properties from sonic tests, *Monograph No. 2*, American Concrete Institute, Detroit, Michigan (1966).

[53] Wongsosaputro, D., *Behavior of plain and high strength concrete*, a report submitted to F.O. Slate at Cornell University, Ithaca, New York (Dec. 1981).

Interferometry in scattered coherent light applied to the analysis of cracking in concrete

6.1 Introduction

The comprehensive understanding of concrete cracking is a designated objective encouraged by the various authorities and organisations at an international level. The two principal motivations in this direction are naturally safety and economy. Conservative design measures or, if necessary, the demolition and reconstruction of a structure are steps preferably avoided. Tacitly this recognises that concrete remains relatively one of the least well-known of construction materials with respect to its cracking behaviour. In addition, the complex mechanisms taking place form a subject of multidisciplinary study where, side by side with theoretical analysis, experimental techniques have a role of utmost importance. In this respect three groups of optical methods can be considered. They are speckle photography, speckle interferometry, and holographic interferometry.

It is accepted from the first that these techniques are not yet widely used in the application foreseen. Their relatively recent arrival, the time necessary for theoretical evaluation and the period for development of efficient equipment are just a few reasons explaining this, as yet, limited usage. For these reasons it has been considered worthwhile to organise this chapter around three themes. Firstly the suitability of the various methods of interferometry in scattered coherent light and the analysis of cracking in concrete are examined. Then the significance of the fringe patterns obtained in the principal optical arrangements is revised. Finally a brief overview of some important applications is given.

6.2 The suitability of interferometric methods in scattered coherent light in the analysis of concrete cracking

Background. Frequently returned to, classical optical methods of the interferometry type have been successfully applied to the analysis of cracking [1–5]. In principal, along with the absence of contact with the specimen, there are two important advantages which give these techniques high performance. These are their good spatial resolution and high sensitivity in the measure of fields of displacement. The two quantities are intrinsically linked to the optical wavelength and in the order of 0.5 μm.

In addition these methods supply information of a continuous nature on extended fields of observation in contrast with that of 'point by point' methods.

The principal handicap of these classical methods lies clearly in their necessity for highly polished surfaces expressed in terms of fractions of the wavelength. However interferometric methods with scattered coherent light avoid this disadvantage whilst retaining all the advantages previously given.

Global characteristics of scattered coherent light interferometry. At the surface of a concrete specimen a crack manifests itself both by the presence of a tiny opening and by the existence of a discontinuity in a field of displacement. The methods of speckle metrology and holographic interferometry simultaneously fulfil two functions: they supply an optical image of the surface and a system of interference fringes covering that image.

The direct detection of the microfissures corresponds to the function of image formation. The distribution of light at the surface of the specimen is immediately affected in the vicinity of the crack which absorbs, diffracts and reflects the incident light in a different way from that of the adjacent regions. For this function the optical systems must possess the highest possible spatial resolution. However, with scattered coherent light, as with incoherent light, the working limits are very narrow. Because of the presence of speckle noise in the image, it is necessary to consider the mean value of intensity in the image. So, for this mean value, the transfer function of the optical system is the same as for incoherent light [6]. The method which consists of increasing the numerical aperture (N.A.) of the system in order to improve the resolution in linear proportion is thus rapidly without real improvement. For a lens free of aberrations the depth of field is inversely proportional to the square of the numerical aperture. In addition, at high numerical apertures the aberrations tend to become dominant. In this situation, the dimension of the field in which the image retains its nominal resolution varies as the inverse of the square or even cube of the numerical aperture depending on the construction of the lens. A twofold gain in the resolution is therefore generally accompanied by a loss of a factor of 16 or 32 in the explorable volume. In electron microscopy the use of very short wavelengths and small numerical apertures enables a very high resolution whilst retaining a useful depth of field [7]. Optically and in the case of a material as rough as concrete, it is reasonable to rely on resolutions of several microns in order to conserve the field advantage although resolutions as good as 0.1 μm have been reported in particular conditions [8].

To the formation of interference fringes corresponds the indirect identification of cracks by the discontinuity disturbance which they provoke in the displacement-deformation field associated with the deformed specimen. In holographic interferometry and speckle metrology these fringes may be lines of equal displacement (u, v, w) or of equal value to the partial derivatives of these components or even combinations of them. A rapid increase in fringe density in a given area followed by the appearance of discontinuities in the fringe pattern (in line or tangential to the fringes) indicates the birth, propagation and line of cracking. For this second function, it is worth noting that:

(1) The crack path is revealed in a point by point manner. The tracing of its route between two consecutive discontinuities introduces the problem of fringe shift, interpolation or multiplication of fringes.

(2) The appearance of the discontinuities in the fringe system depends on the crack mode, the behaviour of the displacement field along the crack and the nature of the displacement components revealed interferometrically. Figure 6.1 shows some cases where the mode is pure, the behaviour of the displacement field simple and the interferometric method exclusively sensitive to the component subtending the largest jump in displacement. Sometimes the fringe discontinuity is not sufficiently well highlighted. In this case, it is usually possible to improve matters by superposing a fictitious rotation, equivalent to a deformation of the specimen, by means of a minute change in the relative inclination of the interfering beams. Given the very powerful interferometric sensitivity it is possible that the jumps caused by minor components might be more interesting to produce. In general neither the mode nor significance of the fringes is pure. Quantitative interpretation is thus not immediate but the qualitative features are not changed.

(3) Considering the subject of resolution, the two functions, interferogram and image formation, appear complimentary. Interferometrically a fringe system of low spatial frequency must be resolved. For the final observation system (eye, TV camera) the resolution requirements are thus considerably relaxed and consequently the information in the fringes is not of an ultimate precision in the establishment of crack geometry. In contrast, however, the unique characteristic of the interferogram is to permit a global detection and instantaneous localization of cracks of a few microns of opening on surfaces ranging from $1\,dm^2$ to $1\,m^2$, provided that the cracks are long enough to cut at least one fringe. On the other hand, considered as an image, the interferogram may be analysed and scanned with a high resolution observation system (microscope) such that the effective resolution is that of the interferometer, that is to say, as previously mentioned, of a few microns. In this way precise information on the crack geometry is obtained although now in a sequential manner as the microscope only sees a fraction of the total field at any one time.

(4) The final basic characteristic stems from the possibility to measure deformations in the neighbourhood of the crack and clearly relates to one of the requirements formulated in fracture mechanics. From the parameters of the fringes it is possible to establish, at least partially, the state and redistribution of strains both in the immediate vicinity of and more distant from the crack. In the case of concrete, recent studies have shown the importance of possessing this sort of information well away from the region where the crack surfaces [9]. Hillerborg's 'fictitious crack model' gives a second example of the value of this property [10–11]. The model is based on the acceptance that, for concrete, there is no clearly defined tip to the crack and the imprecise applicability of formulae expressing fracture mechanics parameters as functions of crack length arise from it. The model thus raises the question of relating one with another, in a zone of cracking, the region of maximum stress and the areas of micro-cracks visible by microscope, then by eye. All are functions to be fulfilled by the proposed methods.

By virtue of their dual action, interferometric techniques using scattered coherent light offer, in one, the possiblity to identify, in a single shot, both the causes — stress concentrations revealed by an accumulation of fringes — and the effects — crack creations in areas where loading is too great and where there are material faults.

Concrete, a suitable material for interferometric methods in coherent light. On a small scale, concrete is a multiphase, heterogeneous material in which the internal structure

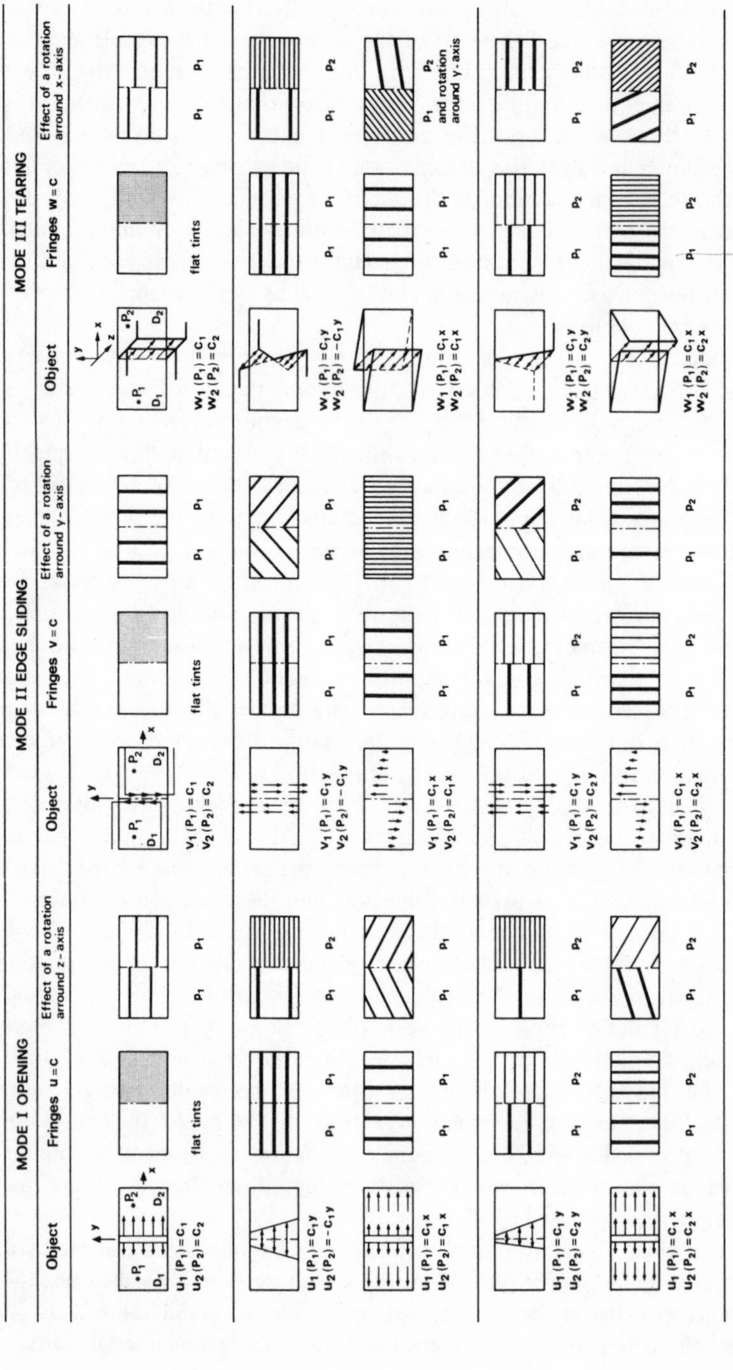

Figure 6.1. Discontinuity aspects in the fringe pattern, for the three modes and different behaviour – constant, linear symmetric, linear nonsymmetric – of components undergoing the preponderant discontinuity.

changes over a very long period. Its behaviour at cracking [12–13] is very hard to understand as a whole mostly because the typical dimensions of its structure cover a very large range.

(1) The field of application of the proposed methods only becomes possible above the molecular level. Nevertheless, indirectly these may sometimes turn out to be useful at this level. With the intention of modelling the phenomenon of shrinkage, Wittmann [14] established the behaviour of Van der Waals forces at short distances, up to 80 Å, and their dependence with respect to the dielectric properties midway between the interacting surfaces. The technique involves measuring the bending of a thin plate under the effect of these forces by classical interferometry. Interpolation techniques to $1/1000^a$ of a fringe developed in speckle and holographic interferometry by Joyeux [15] and Dändliker [16] are of the type to facilitate such fundamental investigations. Excellent references [17–19] are available covering the study of concrete at molecular level with respect to shrinkage, creep and cracking. They contain, among others, the works of Powers, Ishai, Feldman and Sereda, and Wittmann. It should simply be remembered that the models applying at a more coarse level must, necessarily, satisfy the most intimate physical mechanisms.

(2) At the microscopic level, between 0.1 and 1000 μm, numerous flaws are present which will tend to encourage microcracks. These include:

a. the so-called capillary pores inside the C–S–H gel. As, in practice, it is not possible to add exactly the correct quantity of water just to ensure a complete hydration of the cement, this is always saturated. The excess of water occupies small cavities and, on evaporation, leaves a porous structure. The capillary pores, as opposed to gel pores, are in fact the biggest in a continuous statistical distribution. Both the total porosity and distribution of sizes influence the strength of the concrete;
b. preexisting micro-cracks of various origins. Shrinkage, along with the large internal stresses it develops, is an important source of micro-cracking even in a concrete where the phenomenon is inhibited by the presence of aggregates. Bleeding provokes the formation of cracks or zones of weak adhesion underneath large pieces of aggregate or reinforcing. Water, making its way to the surface, has the tendance to become trapped at these points. These cracks lie predominantly in the horizontal plane and introduce a sense of anisotropy. The act of hydration itself creates heat which may introduce cracks of a thermal origin;
c. the interfaces between the hardened cement paste and the inclusions (aggregate, reinforcing). These are generally weak areas and associated with these flaws. At the microscopic level it is the roughness and cleanliness which play an important role in the strength of bond with the paste.

Models of the micro-mechanics of fracture [20–22] use both statistical concepts, to take into account the random distribution of flaws, and criteria of rupture, applied locally and iteratively. They permit, at the highest level achievable, a computer simulation of the behaviour of concrete up until failure in a wide variety of circumstances. However, whether in their checking, calibration or routine use, they need information of experimental origin. This information concerns fault distribution, precracking geometry, preferential modes of crack propagation and their interaction, arresting mechanisms by natural or artificial inclusions, . . . The requirements at this level thus correspond one by one with those which interferometric methods in coherent light are likely to reveal.

(3) At the macroscopic level, concrete may be considered as a two-phase material. That is a homogeneous matrix of hardened cement binding together the aggregates. Macroscopic faults are largely due to a lack of care in production: large pockets of air left trapped after poor compaction, concentrations of large aggregate because of segregation during pouring, long surface cracks in the absence of correct curing or sufficient joints.

At this level, the severity of a crack is expressed by a single parameter such as the stress intensity factor, the energy release rate, the fracture surface energy, in linear fracture mechanics, or the crack opening displacement and J-integral in non-linear mechanics [23]. Without entering into the controversy of whether the critical values of these parameters represent a material property for concrete or not, it should be recognised that it is possible to measure these parameters by way of interferometric methods in coherent light.

It is clear from this section that concrete, because of its very complex structure, demands, from the experimental point of view, a collection of investigation techniques, amongst which interferometric methods in coherent light appear to be quite appropriate, particularly at the microscopic level.

6.3 Significance and quality of fringes obtained with interferometry in scattered coherent light

More than twenty basic optical arrangements of distinct application may be found in speckle photography, speckle interferometry and holographic interferometry. In holographic interferometry, the fringe equations were quickly established in generalised forms which has stimulated the study of particular configurations enabling an easy interpretation of the interferograms. In speckle photography and interferometry the opposite has occurred. It is due to Stetson [24] that not only have all the methods of speckle photography been unified but also that the solution has been shown to come directly from results obtained in holographic interferometry. There now exists a broad agreement in the ways of explaining the significance of fringes from the three groups of methods.

Principal properties of scattered coherent waves. The image or diffraction pattern of a rough surface illuminated with coherent light is heavily speckled. These are spatially random distributions due to multiple interference between contributions arising from a random division of the incident illuminating beam by the object surface [25, 26]. Of particular interest is the case where the speckling obeys a gaussian distribution: if A_r and A_i represent the real and imaginary parts of the speckle amplitude and I the intensity, then [27],

$$p_A(A_r, A_i) = \frac{1}{2\pi \sigma_A^2} \exp - \frac{A_r^2 + A_i^2}{2\sigma_A^2} \tag{6.1a}$$

166

$$p_I(I) = \frac{1}{\langle I \rangle} \exp - \frac{I}{\langle I \rangle}; \qquad I \geqslant 0 \qquad\qquad (6.1b)$$

$$p_I(I) = 0; \qquad I < 0 \qquad\qquad (6.1c)$$

where $2\sigma_A^2$ represents the mean intensity $\langle I \rangle$. For this to be satisfied, it is primarily necessary that the various contributions accumulate in a coherent manner. Concrete, which depolarizes the light, only fulfils this condition if a linear polarizer is introduced between the object and receiver. In addition, the r.m.s. surface roughness must be greater than the wavelength and the correlation length of the roughness small in relation to the dimension of the zone contributing to the speckle amplitude. These last two requirements do not present difficulties in the case of concrete. Supplied with these probability densities the evaluation of much statistical information is made more easy. It should be remembered that the complex amplitude fluctuates in phase equiprobably in a 2π interval and that it is correlated on a small volume of lateral, $\langle p_l \rangle$, and longitudinal, $\langle p_L \rangle$, dimensions given by:

$$\langle p_l \rangle = \frac{\lambda D}{\phi} \qquad\qquad (6.2a)$$

$$\langle p_L \rangle = \frac{8\lambda D^2}{\phi^2} \qquad\qquad (6.2b)$$

where λ is the wavelength, D the distance object-observation plane or the distance lens-image depending on whether one observes the diffraction pattern or the image, ϕ the size of the illuminated zone on the object or alternatively the lens diameter. These represent the mean dimensions of the speckle grains.

What happens to the speckle patterns in the case of small object deformation? According to the concept of homology proposed by Viénot et al. [28], the speckle figures, on a local scale, respond to such a loading in an overall displacement without appreciable structural modification. Thus, each speckle grain represents a datum from which it is sufficient to follow the three dimensional displacement in order to find information on the displacement-deformation field associated with the object surface, as shown in Figure 6.2. Developed mathematically, this concept leads to the formulation of laws of speckle displacement as functions of the displacement-deformation parameters of the object surface [29–33]. A rigorous treatment of this problem is given by Schumann in the [34]. Further mention here will be limited to the schematic presentation of the equations in the form of:

$$\begin{bmatrix} \Omega_x(M) \\ \Omega_y(M) \\ \Omega_z(M) \end{bmatrix} = [C_1] \begin{bmatrix} u(P) \\ v(P) \\ w(P) \end{bmatrix} + [C_2] \begin{bmatrix} R_1(\bar{P}) \\ R_2(\bar{P}) \\ R_3(\bar{P}) \end{bmatrix} + [C_3] \begin{bmatrix} \epsilon_{xx}(P) \\ \gamma_{xy}(P) \\ \epsilon_{yy}(P) \end{bmatrix} \qquad (6.3)$$

where $(\Omega_x, \Omega_y, \Omega_z)$ are the components of speckle translation at the point M of observations; (u, v, w) the displacement components of the center P of the object region which contributes to the illumination in M; (R_1, R_2, R_3), the three local rotations of the surface element do surrounding P; $(\epsilon_{xx}, \gamma_{xy}, \epsilon_{yy})$, the in-plane strain

167

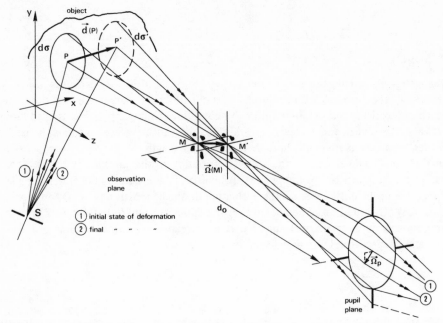

Figure 6.2.   Illustration of the invariance property – on the local scale – of the fine structure of the speckle when the object is deformed.

coefficients at the point P. The matrices C_i [3 × 3] have for elements the terms which are functions of the geometrical parameters of the recording system (Figure 6.2).

The function of the pupil in the observation system is to select the dimensions of the object zone, $d\sigma$, which contribute to the amplitude at the point M under observation. It imposes the dimensions of the correlation volume or mean grain dimensions in accordance with equations (6.2a) and (6.2b).

The displacement field of the speckles is obtained by specialized systems. Three principal procedures are available which involve, for each of the two states of object deformation:

(1) recording in a plane the intensity of the two speckle patterns or observing directly the movements of isolated grains in the microscope. This is thus referred to as speckle photography;

(2) recording in a plane the intensity and phase of speckling. To record the phase of speckling, or rather its evolution, it is necessary to record each time the interference pattern between two speckles, an object and a reference. This leads to the name of speckle interferometry;

(3) providing together the amplitude and speckle phase in volume as takes place in the case of holographic interferometry.

Each procedure will now be examined.

Speckle photography. The principles of speckle photography are given in [35–40]. The method requires the double exposure of a photographic plate to the speckle intensities. That is to say that two positions are recorded for each grain and correspond

168

to the two states of object deformation (Figure 6.3). An essential condition is to be able to identify the two grains and hence the axial translation of the speckles must always stay less than their longitudinal dimension:

$$\Omega_z(M) \ll \langle p_L \rangle \tag{6.4}$$

How can the speckle displacement, 'frozen' in the photographic emulsion, be visualized?

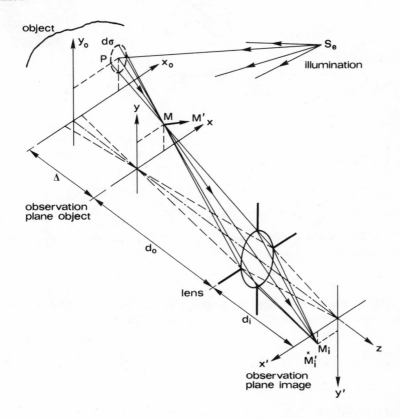

Figure 6.3.   Geometry of the focused ($\Delta = 0$) and defocused ($\Delta \neq 0$) speckle photography.

Point by point filtering leads to the formation of the diffraction pattern of a small area of exposed plate, as shown in Figure 6.4. This image consists of:

(1) a diffraction halo whose mean profile has the appearance of the incoherent transfer function of the recording lens (Figure 6.4a) and thus for diameter ϕ_H:

$$\phi_H = 2 \frac{\lambda_r}{\lambda e} \frac{d_r}{F(1+g)} \tag{6.5}$$

where λ_e and λ_r are the recording and the reconstruction wavelength; d_r, the distance between the plate and the observation screen; F and g, the aperture number and the magnification used;

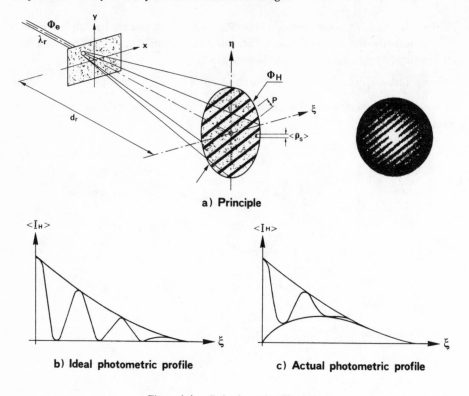

a) Principle

b) Ideal photometric profile   **c) Actual photometric profile**

Figure 6.4.   Point-by-point filtering.

(2) Young's fringes in the direction perpendicular to the displacement vector of the
speckles and whose period, p, is related to the modulus of this vector by:

$$p = \frac{\lambda_r d_r}{(\Omega_x^2 + \Omega_y^2)^{1/2}};$$ (6.6)

(3) secondary speckling, acting as noise, with mean dimension of $\langle p_s \rangle$:

$$\langle p_s \rangle = \lambda_r d_r / \phi_e$$ (6.7)

where ϕ_e represents the diameter of the illuminated region.

The halo diameter and mean dimension of the secondary speckles impose some limits
on the field of displacement of the speckles selected for observation:

$$\langle p_s \rangle < p < \phi_H$$ (6.8)

Ideally, the mean intensity within the diffraction halo will be modulated by sinusoidal
fringes of unit contrast (Figure 6.4b). In reality, however, various factors reduce the
contrast of these fringes (Figure 6.4c) and limit their extension in the halo [41—44].
Amongst these factors are: speckle movement in the plane of the pupil lens, lens
aberrations, the infidelity of the photographic recording and the imperfect overlapping
of the two regions illuminated for restoration.

There exists one notable exception to the rule of sinusoidally profiled fringes. The
presence of a flaw or crack inside a body may be shown by anomalies in the vibratory

behaviour when the body is subjected to periodic excitation. When the speckle is activated by a sinusoidal movement and the experiment performed for a period relatively great in respect to the period of vibration, then the modulation is given by the square of the Bessel function of the first kind and zero order. The appearance of the halo is significantly changed [45].

Alternatively whole field filtering (Figure 6.5) permits visualisation of lines of equal displacement of the speckles in the plane of the plate:

$$\Omega_x(M) = \frac{n\lambda_r f}{u_0} \qquad (6.9a)$$

or

$$\Omega_y(M) = \frac{n\lambda_r f}{v_0} \qquad (6.9b)$$

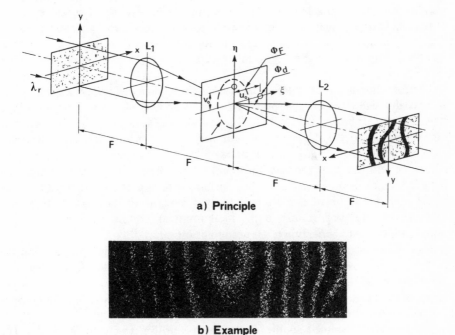

a) Principle

b) Example

Figure 6.5.   Whole-field filtering.

The filter is placed at a distance u_0 from the optical axis along the axis $0x$ or v_0 along axis $0y$. n is an integer and f the focal length of the lenses. This method of visualisation is subject to the same restrictions as the preceding one.

In speckle photography the phases of recording and recovery are distinct and the system does not work in real time. However, two techniques permit an approach to the conditions of real time. In the 'sandwich technique' [46, 47] a series of single exposures is made with a camera. The shots are then recomposed back to back or in a Mach-Zehnder interferometer. The other technique requires the use of new photo-sensitive material with quasi-instantaneous development [48, 49]. The point-by-point analysis of results may now also be carried out automatically [50–52].

In focused speckle photography [35–38], the surface of the object is focused exactly onto the photographic plate. The speckling then moves as if attached to the object surface. In equation (6.3) the fact that the distance between object and observation plane is zero leads to the cancelling out of the majority of matrix elements $[C_i]$. The displacements (Ω_x, Ω_y) of any point $P(x, y)$, on the object surface are given by:

$$\Omega_x(P) = u(P) + w(P)\frac{x}{d_0} \tag{6.10a}$$

$$\Omega_y(P) = v(P) + w(P)\frac{y}{d_0} \tag{6.10b}$$

where d_0 is the distance between the lens and the object, at unit magnification. The influence of out-of-plane displacement is negligible for points near to the optical axis and when the lens characteristics (field angle, focal length) make the approximation $x/d_0 \simeq y/d_0 = 0$ acceptable [53, 54]. An alternative involves simultaneous exposure of two photographic plates by way of two lenses centred on two distinct optical axes. The consequent removal of two series of relations (6.10 a and b) allows resolution of the three components of the displacement vector for every point of the object surface [55].

The recording methods in defocused speckle photography all revolve around several variants although all record the speckle intensity in another plane to the object surface albeit in front or behind (Figure 6.3). The speckle displacement is thus left predominantly sensitive to the slopes of the deformed surface. In one of the arrangements proposed by Gregory, [56], the observation system is focused on the plane lying symmetric with the light source with respect to the plane of the object. In addition, the direction of lighting must be slightly inclined with respect to the normal from the object. For the region of observation exactly symmetric to the source, the speckle displacements are proportional to the local rotations of this intercepted zone. In neighbouring regions correction terms need to be introduced.

In a second arrangement, Gregory, [57], intermingles the lighting source and the centre of the observation system whilst always maintaining the defocusing. The directions of lighting and observation are thus mixed whatever may be the current object point under examination. In consequence the shifts of speckling do not change their significance but only their sensitivity.

Chiang, [58], proposed another arrangement sensitive mainly to the slopes. The object is illuminated by a parallel beam inclined at angle ϑ to the normal. In the absence of rigid body displacement and when the in-plane displacement is small with respect to the out-of-plane displacement, then the equations (6.3) reduce to:

$$\Omega_x(M) = \pm\Delta\left[(1 + \cos\vartheta)\frac{\partial w}{\partial x}(P) + \sin\vartheta\frac{\partial u}{\partial x}(P)\right] \tag{6.11a}$$

$$\Omega_y(M) = \pm\Delta\left[(1 + \cos\vartheta)\frac{\partial w}{\partial x}(P) + \sin\vartheta\frac{\partial u}{\partial y}(P)\right] \tag{6.11b}$$

where Δ is the amount of defocus.

172

These various arrangements, of variable sensitivity or significance, are mostly suited to particular problems such as bending of thin plates. They are more thoroughly described in [33].

The general method was advocated by Stetson, [59], and requires the exposure of at least five focused and defocused specklegrams of which some are set up in tandem in order not to complicate the arrangement. Although delicate to set up this method is the only one which supplies a complete displacement-deformation field of the surface being studied.

The white light speckle photography method is identical to that of focused speckle photography, [60–62] except for the light source. The granular structure, serving to record the position and then follow the displacements in all areas of the specimen, has to be artificially created. This may be by either covering the surface with a special coating or highlighting the surface irregularities with sufficient contrast. The principal advantage of the system lies in the possibility of illuminating large surfaces at minimum cost. The limitations imposed by the relationship (6.8) rest basically unchanged. The advantages of defocusing, as with coherent light, no longer exist as speckling from white light only occurs in a narrow range dependent on the depth of field of the lens.

Speckle interferometry. The principles of this method can be found in [63–66]. The changing of the interference pattern between the specimen speckling and a control encodes the deformation of the specimen. These latter speckles serve as a reference and may or may not be affected by the deformation of the specimen, (Figure 6.6).

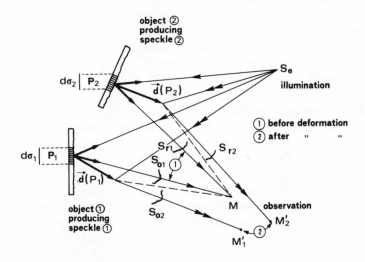

Figure 6.6.   Schematic of speckle interferometry.

If the lateral and longitudinal dimensions of each speckle are chosen smaller than their displacement components when the object deforms then, contrary to speckle photography, the speckle movement will not be noticed with the speckles resting mixed:

$$(\Omega_{x0}^2 + \Omega_{y0}^2)^{1/2} \ll \langle \rho_{l0} \rangle; \qquad \Omega_{z0} \ll \langle \rho_{L0} \rangle \tag{6.12a}$$

$$(\Omega_{xr}^2 + \Omega_{yr}^2)^{1/2} \ll \langle \rho_{lr} \rangle; \qquad \Omega_{zr} \ll \langle \rho_{Lr} \rangle \tag{6.12b}$$

The subscripts 0 and r refer to the specimen and reference speckle respectively.

The intensity of a given grain changes as a function of the variation of phase difference between the two speckles which is provoked by the deformation of the specimen. In regions where the variation of the difference of optical path between the object speckle and reference speckle is an integer multiple of the wavelength, the resultant final speckle is identical to the resultant initial speckle. For a variation of half λ difference the two interfering speckles change their phase relationship by π. In other words, the sum of the resultant initial and final speckles represents exactly an incoherent addition of the two images. The fringes are thus signified by:

$$\Omega_{zr}(M) - \Omega_{z0}(M) = n\lambda \tag{6.13}$$

where Ω_{zr} and Ω_{z0} are respectively the longitudinal displacements of the specimen and reference speckles. These two components may be found by using equation (6.3). The techniques of visualisation are comparable with those in additive or multiplicative moiré. With double exposure the two initial and final speckles are superposed on the same photographic plate before development. The regions undergoing an integer wavelength variation have characteristic features of a granular image with unit contrast whilst the regions of half λ variation possess a granular structure of weaker contrast, characteristic of incoherent addition. However, if the photographic process is linear in intensity, the mean intensity transmittance in the two regions is the same and makes observation difficult. A non linear photographic process gives some contrast to the average intensities of both regions. In the mask technique the final resultant speckle is viewed through the repositioned, developed negative of the initial resultant speckle. The regions of integer λ variation appear as dark fringes spotted with a few brilliant individual speckles whereas the regions of $\lambda/2$ variations keep their typical speckle appearance. A non linear record of the initial state reinforces, more than previously, the contrast between the average intensities of the two regions. The mask technique is intrinsically superior to that of double exposure. It functions in real time, permits compensation of rigid body movement by successive returns to the initial plate and gives fringes of better quality.

A more elaborate form of visualisation makes use of an 'Electronic Speckle Pattern Interferometer' [64–66]. The interference between the specimen and reference speckles is picked up by a camera of a video system. The instrument carries out a chain of operations: recording on magnetic tape, playback, mixing, filtering and contrast improvement. The time required to record the initial interference pattern is thus very short. However, as the video tube has a weak spatial resolution, it is this which governs the mean dimension of the speckle and no longer the relationships of equations (6.12a and b). In conequence the numerical apertures of the objective lenses are weak and the speckles larger than otherwise required.

It is difficult to assess the photometric profile of fringes of speckle interferometry due to the non linear processes, usually difficult to monitor, which are used for the recording and visualisation. Consequently all attempts at interpolation by microdensometric methods is fraught with problems. The fringe contrast, may theoretically, lead towards one, but only by very heavy exposure and resulting ultimately with only a poor yield on examination of the plates. As with speckle photography, both the speckle shift in the pupil plane and the imperfect superposition of the grains in the image plane are the principal factors contributing to the reduction of fringe visibility [67, 68]. Finally, because of the conditions imposed by the relationships, equations (6.12a and b), the interferograms are relatively 'noisy'.

The significance of the fringes in several special arrangements will be discussed.

Speckle interferometry can be adapted to measure the out-of-plane displacements. A recording lens is aimed normally at the specimen surface which is focused onto the photographic plate (Figure 6.7a). The reference speckle, independent of the object deformation, is, in reality, a plane or spherical wave propagating itself in the same direction as the speckling coming from the object [69]. If, in addition, the object is illuminated at zero incidence the equation (6.13) reduces to:

$$w(P) = \frac{n\lambda}{2\left(1 + \dfrac{\tan^2\phi}{4}\right)} \tag{6.14}$$

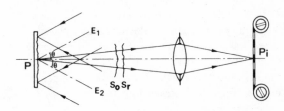

a. Speckle interferometer adapted to out-of-plane displacements measurement.

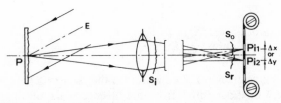

b. Speckle interferometer adapted to in-plane displacements measurement.

c. Speckle shearing interferometer.

Figure 6.7.   Speckle interferometry setup. a. Speckle interferometer adapted to out-of-plane displacements measurement. b. Speckle interferometer adapted to in-plane displacements measurement. c. Speckle shearing interferometer.

where ϕ is the field angle of the point, P, under observation. Once again, this particular geometry considerably simplifies equation (6.3). For the points, P, remote from the optical axis, $\tan^2\phi$ may no longer be negligible so that the sensitivity is not constant across the full field. This sensitivity is otherwise excellent and may be around $0.25\,\mu m$ per fringe. The same fringe equation is obtained from an arrangement where the two mirrors of a Michelson interferometer are replaced by two scattering surfaces [63]. One is fixed to supply the reference speckle and the other is the specimen surface observed normally by a lens focused on its surface.

The measurement of in-plane displacements, is accomplished by using a symmetric double illumination of the specimen. Object and reference speckles are created. They are mixed in their propagation and both influenced by the deformation of the specimen (Figure 6.7b), [70]. The variation of the difference of path length between the two speckles is thus made sensitive to the displacement component in the plane of the specimen and parallel to the directions of specimen illumination. The fringes of equation:

$$u(P) = \frac{n\lambda}{2\sin\vartheta} \tag{6.15a}$$

or

$$v(P) = \frac{n\lambda}{2\sin\vartheta} \tag{6.15b}$$

are derived, where ϑ is the angle of incidence of the illuminating beams. In collimated parallel light the method does not suffer a change of sensitivity across a field. A variation of this method involves replacing the double illumination with a symmetric double observation by means of a pupil pierced by two holes [71].

In speckle shearing interferometry [72, 73], the photographic plate receives two interfering speckles shifted laterally but originating from the same object. The operation is achieved by means of a classical interferometer, placed between object and image, before or after the lens (Figure 6.7c). When the interferometer is adjusted to introduce a shift, Δx or Δy, between the two speckles then the variation of optical path is a linear combination of the partial derivatives of the components of the object displacement in the two direction. Respecting the equations (6.3) and (6.13), the fringe equation may be written as:

$$\sin\vartheta\,\frac{\partial u}{\partial x}(P) + (1 + \cos\vartheta)\frac{\partial w}{\partial x}(P) = \frac{n\lambda}{\Delta x} \tag{6.16a}$$

or

$$\sin\vartheta\,\frac{\partial v}{\partial y}(P) + (1 + \cos\vartheta)\frac{\partial w}{\partial y}(P) = \frac{n\lambda}{\Delta y} \tag{6.16b}$$

with ϑ being the incident angle of illumination in the plane $(x0z)$ or $(y0z)$.

In general these methods suffer from the inconvenience of a limited object field and away from the optical axis correction terms are introduced [33, 74, 75].

By choosing a zero angle of incidence the influence of in-plane deformation is eliminated. With the simultaneous exposure of two plates for the two illumination directions, it is possible to recover the two components ($\partial u/\partial x$ and $\partial w/\partial x$) by calculation, [76].

Holographic interferometry. Referring to the principles in [34, 77, 78], two speckles associated with two states of object deformation are compared in amplitude, phase and volume by way of a holographic record. Accepting that the procedure of holography is well-known, [79, 80], the only feature which needs recalling is that it provides a means of simultaneously reconstructing two non concomitant waves, as shown in Figure 6.2. By referring to the properties of invariance and correlation of the two speckles, the interference between the two waves will give microscopic fringes only in the regions of space where they are shifted by a quantity less than the dimensions of the volume of correlation:

$$(\Omega_x^2(M) + \Omega_y^2(M))^{1/2} \ll \langle \rho_l(M) \rangle \tag{6.17a}$$

$$\Omega_z(M) \ll \langle \rho_L(M) \rangle \tag{6.17b}$$

These are the same relationships as in speckle interferometry. They lead to a different interpretation and introduce the notion of fringe localization produced holographically [28, 34, 81−86]. As it is now possible to explore the field of potential fringes in volume, it is no longer worthwhile imposing, on this basis, a large granulation on the speckle. The apertures used in holographic interferometry are optimal as they may be adjusted later, during the reconstruction, as dictated by the requirements of equations (6.17a and b) and as a consequence are much bigger than in speckle interferometry. The quality of the interferograms is thus significantly improved. In a localization area, where the equations (6.17a and b) are satisfied, the fringes may be simply described as lines of equal height difference between the two speckles:

$$\Omega_{z0}(M) + \left[\frac{\Omega_{xp}}{d_0}\right] x + \left[\frac{\Omega_{yp}}{d_0}\right] y = n\lambda \tag{6.18}$$

Here Ω_{z0} is the longitudinal translation between the two speckles at point M and in its vicinity. The coefficients of x and y represent the relative inclination of the two waves expressed as the ratio between the components of translation in the plane of the pupil and the distance, d_0, between the pupil and the plane of observation (Figure 6.2). The relationship (6.18) clarified by means of relation (6.3) reveals the complex liaison between the fringe characteristics and the displacement-deformation parameters of the object surface. These latter parameters influence the period, direction, order, whether integer or fractional, and the localization of the fringes under consideration.

Confronted with this situation three options are open:

(1) to simultaneously record several holograms in order to prepare a sufficient number of independent equations of the form (6.18) as is done in tandem speckle photography and explained in the references [87−91];

(2) to determine lines or surfaces of localization, by photometric analysis of the fringe contrast, which enables a direct determination of the strain coefficients of the object surface [92−93]; and

(3) to accept partial, but simply obtainable information, taken from particular optical arrangements, with the intention of considerably simplifying the relationship (6.18), [94–101]. This approach will be taken up later.

In comparison with methods of speckle metrology the visualisation of the interferograms is immediate. By double exposure, with the help of a continuous or pulsed laser, the hologram is successively exposed to the initial and final speckles. At restoration, the two waves are simultaneously reproduced and they interfere under the conditions previously described. The fringes may be viewed directly (eye, TV camera) in the same way as the restoration of a single exposure hologram.

In real time, the hologram is exposed once and records a wave corresponding to the initial state of deformation. When developed and repositioned exactly the wave recorded is able to interfere at each instant with the wave coming from the object. Holographic cameras using photothermoplastic material [102] mark a significant development over traditional plates. With dry, automatic, in-situ development in a few seconds and, being both cheap and reusable, this type of hologram considerably increases possibilities of application.

With real time it is prudent to incorporate micrometric supports for the hologram, reference and light source to permit the application of compensation techniques [103]. These techniques are intended to ensure the best possible superposition of two speckles interfering in any given plane of observation. Whilst useful in the suppression of the rigid body displacements of the object, they also contribute to an improvement in the interferogram contrast and reduce the speckle noise. They make the use of large apertures effective and worthwhile. The quality of the holographic interferograms is thus generally very good. The fringes, of sinusoidal shape, have a contrast tending to unity. Note that, in the case of harmonic vibrations of the object and, when the exposure is of time-average type, the intensity modulation is given by the square of a zero order Bessel function [104]. The limitations of holographic interferometry compared with those found in speckle metrology arise more because of the great rigidity necessary in the optical arrangement than because of the restriction on the quality of the interferograms.

Holographic set-up can be used for measuring the out-of-plane displacements. The directions of lighting and observation are symmetric with respect to the normal from the specimen surface and, at the limit, coincident. If, in addition, the following conditions are fulfilled:

(1) collimated lighting of the object;
(2) the pupil of the observation system placed at infinity; and
(3) observation of the interferogram at the object surface.

Then the lines of equation (6.18) reveal themselves as fringes of equal out-of-plane displacement:

$$w(P) = \frac{n\lambda}{2\cos\vartheta} \tag{6.19}$$

The compensation techniques enable localization of the fringes at the object surface without loss of aperture and, as required, allow amplification or reduction of the fringe density. The maximum sensitivity is around $0.25\,\mu m$ per fringe.

The measurement of in-plane displacements by holography is discussed in [105–109]. The object is illuminated by two plane waves lying symmetrically about the normal to the surface. The arrangement differs from that of Leendertz only by the presence of a reference wave superposed on the image of the object (Figure 6.7b). The in-plane displacement, that is in the direction containing the two beams, is visualized as a beat or moiré between the two interferograms due to each illuminating beam. To obtain the moiré two conditions must be fulfilled:

(1)   each interferogram must be well localized in the image plane; and
(2)   for good quality the density of the interferograms must be increased by addition of a fictitious deformation state of the sort inducing sufficient carrier fringes (interferogram) which participate to form a beat (moiré fringe).

The local spatial frequency of the moiré is the difference of the spatial frequencies of the two interferograms in the region under consideration. Each interferogram shows the fringes as:

$$[\sin \vartheta] \, u(P) + [1 + \cos \vartheta] \, w(P) + f_r(P) \; = \; n_1 \lambda \qquad (6.20a)$$

$$[-\sin \vartheta] \, u(P) + [1 + \cos \vartheta] \, w(P) + f_r(P) \; = \; n_2 \lambda \qquad (6.20b)$$

where $f_r(P)$ represents the effect of the residual deformation-displacement parameters of the object as well as the effect of the fringe addition. The moiré thus appears as a family of lines given by the difference of the two preceding equations:

$$u(p) \; = \; \frac{n\lambda}{2 \sin \vartheta} \qquad (6.21a)$$

or

$$v(P) \; = \; \frac{n\lambda}{2 \sin \vartheta} \qquad (6.21b)$$

for lighting directions in the $(x0z)$ plane or in the $(y0z)$ plane respectively.

As the contrast of the moiré fringes is only a fraction of the interferogram contrast it is clearly important that the latter have a high contrast [110]. The micrometric compensation adjustments are only efficiently operated in real time. The visualisation of the moiré on a TV monitor greatly eases the adjustments by an instant check on the effect of each manipulation and ensures a rapid estimation of the moiré image by a process of trial and error.

From the point of view of sensitivity and the nature of the measured components, the two preceding examples have their exact twin in speckle interferometry. So why resort to a holographic method recognised as more delicate? It is the quality and flexibility of the system which justifies the additional precautions as may be shown by a few examples.

The interventions possible in holography are much more convenient as the reference wave offers very high degrees of freedom unknown in speckle metrology. This reference wave is invaluable in heterodyne operations [16, 111], optical derivation of interferograms [101] and fringe multiplication [112].

However the foremost advantage is, without doubt, the ability of holography to better combine the two functions of image formation and the formation of interference fringes. In effect the combination of hologram and microscope, or holographic microscopy, [113–117], overcomes, to a very large extent, the weaknesses of poor depth of field common to all high resolution optical systems. It is thus possible to examine, both sequentially and retrospectively with a constantly high resolution, broad three dimensional images related to a unique event. This property is extremely valuable in the analysis of cracking in concrete.

Finally, rather than highlighting the differences between the methods of coherent light interferometry it is perhaps fairer to consider their complementary features and compatibility. In this respect it is possible to combine holographic interferometry and speckle photography whether focused or not. Thus a camera placed behind a hologram records, in real time, a doubly exposed specklegram formed partly by the hologram alone and then by the illuminated object after cutting the reference wave.

The similarity of the principle and a common core of equipment equally lend favour to this compatibility.

6.4 Comparative summary

In reviewing the many aspects covering the analysis of concrete cracking three groups of methods were presented. The summary table (Table 6.1) gives the significance of the fringes and an estimate of the ranges of measurement and sensitivity in each case. The estimation is in one part empirical, in another based on the practical consequences of fundamental limitations (speckle noise, fringe quality) and inspired by several references [101, 118]. The numerical values are expressed for a wavelength of $0.5\,\mu m$ and using an interpolation of 1/5th of a fringe which corresponds, approximately, to a visual mode of assessment. In the case of particular assistance the range and sensitivity are increased proportionally to the gain achieved in the interpolation factor. In this way, a $2\,\text{Å}$ sensitivity may be achieved in holographic, out-of-plane interferometry for a 1/1000 fringe interpolation obtained by heterodyne techniques. Both the systems of collection and digitalisation of images, as well as the new photosensitive materials, simplify the stage of quantitative analysis of results. The combining of these systems with a computer leads, as a whole, to a rapid, high performance means of analysis.

6.5 Application of interferometric methods in scattered coherent light to the analysis of concrete cracking

The analysis of cracking makes use of work published in the general field of non-destructive testing [39, 119–121] and related disciplines. The following paragraphs present and overview of the applications of methods of speckle photography and both speckle and holographic interferometry to the fracture mechanics of concrete. Whenever it seems possible that a transfer may be made to the case of concrete, the applications used with other materials are also mentioned.

Speckle photography. Reference [122] is directed at the detection and localization of cracks. A concrete test piece, $1000 \times 100 \times 50\,\text{mm}^3$, reinforced with a steel rod, is

Table 6.1 Comparative summary of the described methods

Type of methods	Measurable components	Parasite components	Elimination	Range	Sensitivity (based on 1/5 of a fringe)	Model size
Focused Speckle Photography	u and v	w	possible*	$5-500\ \mu m$	$1\ \mu m$	$1\ m^2$
Defocused Speckle Photography	$\dfrac{\partial w}{\partial x}$ and $\dfrac{\partial w}{\partial y}$	u, v, w	possible*	$10^{-4}-5.10^{-3}\ rd$	$2.10^{-5}\ rd$	$0.25\ m^2$
White Light Speckle Photography	u and v	w	possible*	$20-500\ \mu m$	$5\ \mu m$	$5\ m^2$
Out-of-plane SI	w	u and v	yes**	$0.25-10\ \mu m$	$0.05\ \mu m$	$0.25\ m^2$
In-plane SI	u or v	−	−	$0.5-20\ \mu m$	$0.1\ \mu m$	$0.25\ m^2$
Shearing SI	$\left[\dfrac{\partial u}{\partial x}\text{ or }\dfrac{\partial w}{\partial x}\right]$ or $\left[\dfrac{\partial v}{\partial y}\text{ or }\dfrac{\partial w}{\partial y}\right]$	vice versa	possible*	$2.10^{-5}-8.10^{-4}\ rd$ or $10^2-10^3\ \mu strain$	$4.10^{-6}\ rd$ or $20\ \mu strain$	$0.25\ m^2$
Out-of-plane HI	w	u, v	yes**,***	$0.25-20\ \mu m$	$0.05\ \mu m$	$0.25-1\ m^2$
In-plane HI	u or v	w	yes***	$0.5-20\ \mu m$	$0.1\ \mu m$	$0.25\ m^2$

*Analysis of several simultaneously recorded specklegrams or interferograms.
**Collimated illumination, normal to the object surface and the aperture of the observation system at infinity.
***Compensation.

put in three point bending. The point by point analysis of focused specklegrams isolated the cracks to a $1\,dm^2$ area at the centre of the specimen for various load increments. This involved locating areas of the image showing a jump in the period and orientation of the Young's fringes. These observations correspond to those made by microscope. On the crack itself a significant loss of fringe contrast is noticed.

The quantitative aspect is discussed in the works of Luxmoore et al. [123, 124]. The stress intensity factor may be determined, by Westergaard solution, knowing the field of displacement at the crack tip. The authors demonstrate that focused speckle photography fulfils the conditions required by this solution, that is, a sensitivity better than $1\,\mu m$ in the displacement components of the resolved points no more than $1\,mm$ distant. A comparison between components measured by point by point filtering and calculated with a theoretical K_I value shows a very good agreement for a brittle plastic.

To assess the applicability of the Westergaard solution to dynamic problems, Trafalian et al. [125] use a pulsed ruby laser. Whole field fringes and fringes of point by point filtering are exploited successively. The first serve to localize the critical zones from whence a more quantitative analysis is made by the second.

De Backer [126], working on a cantilevered concrete beam, detected cracks of less than $1\,\mu m$ opening by means of point by point filtering, the total surface examined being $180 \times 240\,mm^2$. Under increasing load crack opening as small as $0.35\,\mu m$ was measured by introducing, between the two exposures, a known, fictitious field of displacement.

It may be preferable to detect a crack by the discontinuities in the fringes of equal slope even if the significance of the fringes is not pure. Methods of defocused speckle photography are mentioned in the works of Ennos, [127], and Chiang, [128]. Gregory, [57], made extensive use of focused and defocused arrangements for the measure of crack opening displacement and the study of cracking in pressure vessels and aeronautical structures.

Finally white light speckle was used by Chiang to determine the stress intensity factor and crack opening displacement of a plexiglass compact tension specimen, [129]. The specimen is made opaque by a paint creating artificial speckles and, according to the author, the techniques is applicable to virtually all construction materials.

Speckle interferometry. In the author's knowledge no specific examples exist for concrete cracking even though interferometers adapted to the measurement of in- and out-of-plane displacements may be frequently used in nondestructive testing, [130–131]. Speckle interferometry, nevertheless, counts as one of its assets one of the most powerful methods of crack investigation, due to Barker and Fourney [132], which permits a three dimensional assessment of the stress intensity factor, yet applicable to transparent objects.

On a general scale, Hung, [73], showed that, using shearing speckle interferometry, it is often easier to correlate a structural flaw with some anomalies in the strain than with anomalies in the field of displacement.

Holographic interferometry. The works of Dudderar, [133–134], and Vest, [135–136], are to be distinguished as pioneering work in this field.

For concrete, Light and Luxmoore [137–138] report on experiments made with double exposure on cubes and cylinders in compression. The arrangement is sensitive to out-of-plane components of displacement. It applies to the establishment of surface crack geometry, by successive load increments, up until rupture. The influence of the adhesion to the bearing plates and the centralisation of the load is clearly shown in the interferograms. The naked eye is unable to see a crack until failure although it easily sees the discontinuities in the fringes.

Stroeven et al., [139, 140], work on different compression specimens using arrangements sensitive to out-of-phase displacements which may be either by double exposure or in real time. The aim is to detect and follow the propagation of cracks by studying the most usual zones for creation and growth and measuring crack length as a function of crack opening and load. In real time the loads provoking sudden crack growth are easily located with precision.

Under the same conditions as speckle photography Pflug, [122], analysed the propagation of real cracks in an unnotched concrete beam loaded in three point bending. The two arrangements sensitive to in- and out-of-plane displacements are used successively and, in this case, in real time. The maximum applied load never exceeds half of the rupture load. By virtue of the compensation adjustments, many cycles of loading/unloading are possible with the same hologram and with the corresponding interferograms recorded on magnetic tape, (Figure 6.8). The crack does not increase its length linearly as a function of the loading. The geometry of the strained zone and development at unloading are clearly seen. For in-plane displacement the discontinuities are clearly revealed by the slope of the fringes, (Figure 6.9), which thus presents a symmetry of behaviour.

The example given in Figure 6.10 presents the initial results obtained in the study of crack formation in small, cylindrical cement-paste specimens under the effects of shrinkage*. The first crack appeared a few minutes after the start of drying and it developed in sudden jerks often preceded by fringe oscillation.

In associated domains of study, [141] reports holographic work on the bond forces between concrete and steel reinforcing bars. It is possible to distinguish between liaisons of a chemical adhesive nature and friction. The publications of Spetzler [142] and Holloway [143] refer to the fracture mechanics of rocks, with the latter, in addition, using a pulsed ruby laser for the dynamic problems of crack propagation under explosive charges.

It should be added that numerous quantitative results have been obtained by holographic interferometry for materials other than concrete (i.e. dimensions of the plastic zone, static and dynamic stress intensity factors, energy release rate, effective crack length, . . .). Examples, representative of this range of work, are the steel specimens of Cadoret, [144], and Meyer, [145], the aluminium by Dudderar, [133], homalite 100 by Holloway, [146], and Rossmanith, [147], PMMA by Dudderar, [148], polystyrene by Krenz [149], and so on.

It is equally important to realise that holography has found, through the works of Marom et al., [150, 151], and Ohlson, [152], a very interesting application in the

*Swiss National Research contract awarded to Laboratories of Construction Materials and of Stress Analysis (Professors F.H. Wittmann and L. Pflug), Swiss Federal Institute of Technology, Lausanne, September 1981.

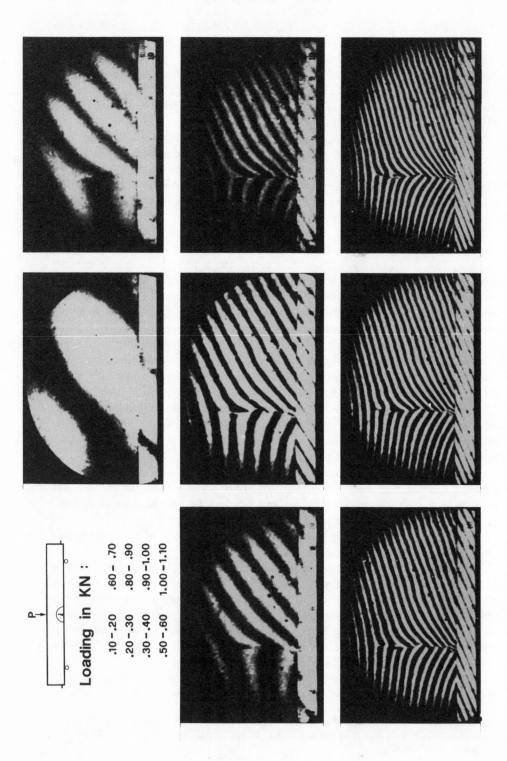

Loading in KN :

.10 – .20   .60 – .70
.20 – .30   .80 – .90
.30 – .40   .90 – 1.00
.50 – .60   1.00 – 1.10

Figure 6.8.   Crack propagation in a concrete beam in three-point bending using holographic interferometry. Evolution of the out-of-plane displacement fringes with increased loading.

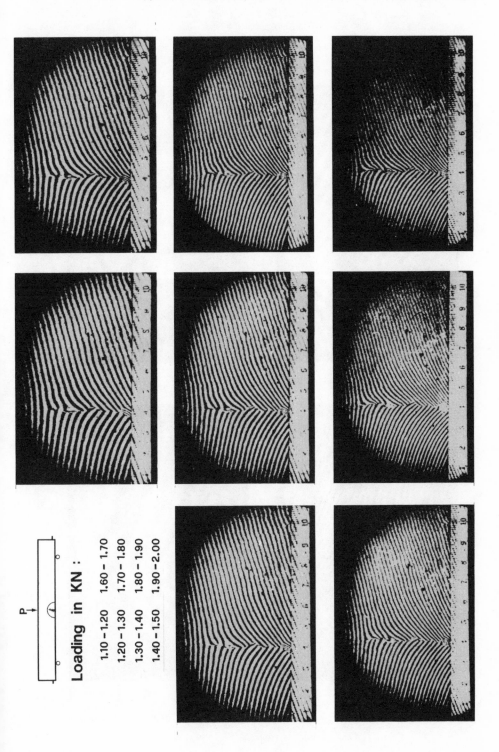

Loading in KN :

1.10 – 1.20   1.60 – 1.70
1.20 – 1.30   1.70 – 1.80
1.30 – 1.40   1.80 – 1.90
1.40 – 1.50   1.90 – 2.00

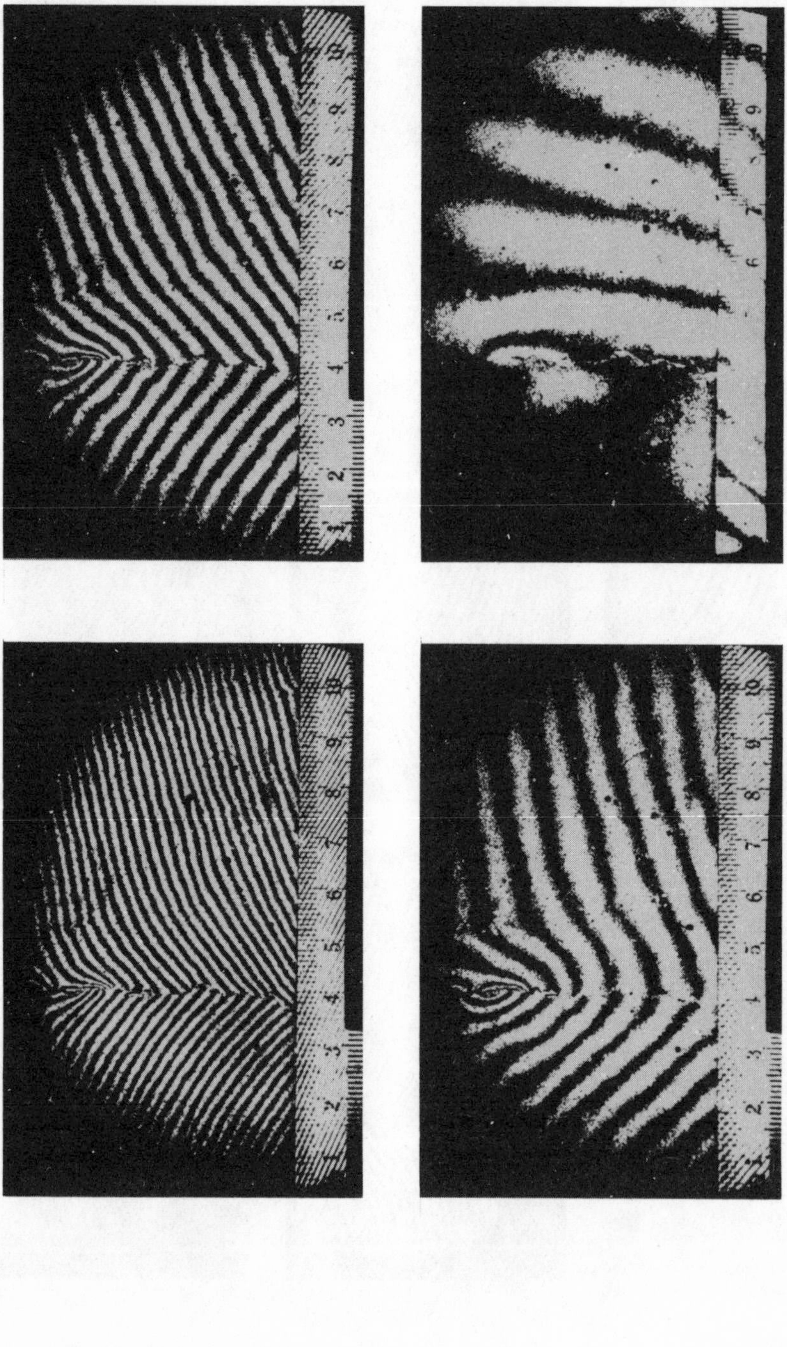

Figure 6.8.   Continued. Evolution during unloading.

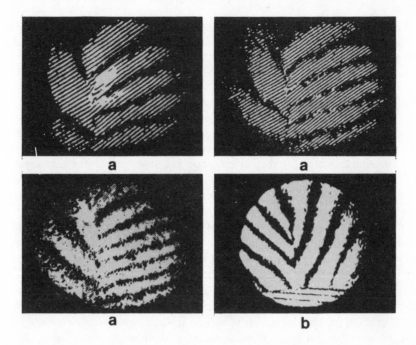

Figure 6.9.   Crack propagation in a concrete beam in three-point bending using holographic interferometry. In-plane displacement fringes obtained (a) on the TV monitor, (b) after optical filtering.

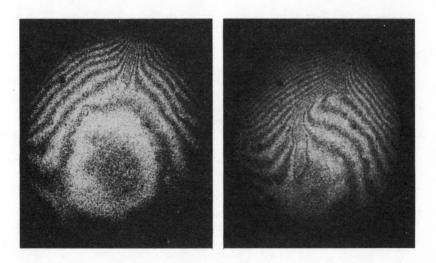

Figure 6.10.   Crack formation in the upper base of a cylindrical cement-paste specimen under the effects of shrinkage, using out-of-plane holographic interferometry.

study of fatigue cracking based on a measure of correlation. Thus although strictly not being an interferometric technique it is a supplementary illustration of the flexibility of the holographic process.

6.6 Conclusion

It cannot be denied that the interferometric methods in scattered coherent light have, as a whole, made available specific services for the analysis of cracking in concrete. The characteristics of spatial resolution and sensitivity in the measurement of the displacement-deformation field place their use principally at the microscopic level which coincides with the development of models in the domain of the materials engineer.

A more frequent and widespread use of these methods is foreseen in the future for the following reasons. The mechanisms of fringe formation, their significance and the quality of the interferograms are well understood at the theoretical level. The equipment is continually being refined with respect to the power achieved and a greater fidelity of the lasers as well as an increasingly more fully automatized system of recording and analysis of results. The outlook, on the side of the mathematical modelling of concrete cracking behaviour at various levels is equally very encouraging. The degree of specialisation reached with respect to both experimental and theoretical methods is now very high. For this reason the future progress may be better encouraged by the formation of interdisciplinary teams of specialists.

References

[1]   Packman, P.F., Role of interferometry in fracture studies, *Experimental Techniques in Fracture Mechanics – 2*, edited by A.S. Kobayashi, Iowa State University Press, Ames, Iowa, pp. 59–83 (1975).

[2]   Underwood, J.H. and Kendall, D.P., Measurement of microscopic plastic-strain distributions in the region of a crack tip, *Experimental Mechanics*, 9, pp. 296–304 (1969).

[3]   Marci, G. and Packman, P.F., The effects of the plastic wake zone on the conditions for fatigue crack propagation, *International Journal of Fracture*, 16, pp. 133–153 (1980).

[4]   Bar-Tikva, D., Grandt, A.F. Jr. and Palazotto, A.N., An experimental weight function for stress intensity factor calibrations, *Experimental Mechanics*, 21, pp. 371–378 (1981).

[5]   Sharpe, W.N.Jr., Applications of the interferometric strain/displacement gage, *Optical Engineering*, 21, pp. 483–488 (1982).

[6]   Lowenthal, S. and Arsenault, H., Image formation for coherent diffuse objects: statistical properties, *Journal of the Optical Society of America*, 60, pp. 1478–1483 (1970).

[7]   Mindess, S. and Diamond, S., A preliminary SEM study of crack propagation in mortar, *Cement and Concrete Research*, 10, pp. 509–519 (1980).

[8]   Higgins, D.D. and Bailey, J.E., A microstructural investigation of the failure behaviour of cement paste, *Proceedings of a Conference at University of Sheffield, Hydraulic Cement Pastes: Their Structure and Properties, Cement and Concrete Association*, pp. 283–296 (1976).

[9]   Sok, C., Baron, J. and François, D., Mécanique de la rupture appliquée au béton hydraulique, *Cement and Concrete Research*, 9, pp. 641–648 (1979).

[10] Hillerborg, A., Analysis of fracture by means of the fictitious crack model, particularly for fibre reinforced concrete, *International Journal of Cement Composites*, 2, pp. 177–184 (1980).

[11] Hillerborg, A. and Petersson, P.E., Fracture mechanics calculations, test methods and results for concrete and similar materials, *Proceedings of the 5th International Conference on Fracture, Cannes, Advances in Fracture Research, volume 4*, Pergamon Press, pp. 1515–1522 (1981).

[12] Les fissurations des bétons, *Annales de l'Institut Technique du Bâtiment et des Travaux Publics, série Béton no. 202*, no. 398 (1981).

[13] Fracture Mechanics of Concrete, edited by F.H. Wittmann, Elsevier Scientific Publishing Company (1982).

[14] Splittgerber, H. and Wittmann, F., Einfluss adsorbierter Wasserfilme auf die Van der Waals Kraft zwischen Quartzglasoberflächen, *Surface Science*, 41, pp. 504–514 (1974).

[15] Joyeux, D., Real time measurement of very small transverse displacements of diffuse objects by random moiré. 1: Theory. 2: Experiments, *Applied Optics*, 15, pp. 1241–1255 (1976).

[16] Dändliker, R., Heterodyne holographic interferometry, *Progress in Optics, volume XVII*, edited by E. Wolf, North-Holland Publishing Company, pp. 1–84 (1980).

[17] Soroka, I., Portland Cement Paste and Concrete, The Macmillan Press Ltd. (1979).

[18] Proceedings of the International Conference Structure of concrete and its behaviour under load, London 1965, Cement and Concrete Association, edited by A.E. Brooks and K. Newman, (1968).

[19] Wittmann, F.H., Structure of concrete with respect to crack formation, *Fracture Mechanics of Concrete*, edited by F.H. Wittmann, Elsevier Scientific Publishing Company, chapter 2, (1982).

[20] Zaitsev, Y.B. and Wittmann, F.H., Simulation of crack propagation and failure of concrete, *Materials and Structures*, 14, pp. 357–365 (1981).

[21] Wittmann, F.H., Mechanisms and mechanics of fracture of concrete, *Proceedings of the 5th International Conference on Fracture, Cannes, Advances in Fracture Research, volume 4*, Pergamon Press, pp. 1467–1487 (1981).

[22] Mihashi, H., A stochastic theory for fracture of concrete, *Fracture Mechanics of Concrete*, edited by F.H. Wittmann, Elsevier Scientific Publishing Company, chapter 4.3, (1982).

[23] Ziegeldorf, S., Fracture mechanics parameters of hardened cement paste, aggregates and interfaces, *Fracture Mechanics of Concrete*, edited by F.H. Wittmann, Elsevier Scientific Publishing Company, chapter 5.1, (1982).

[24] Stetson, K.A., Problem of defocusing in speckle photography, its connection to hologram interferometry and its solution, *Journal of the Optical Society of America*, 66, pp. 1267–1270 (1976).

[25] Laser Speckle and Related Phenomena, *Topics in Applied Physics, volume 9*, edited by J.C. Dainty, Springer-Verlag (1975).

[26] Françon, M., Granularité Laser. Speckle. Applications en Optique, Masson (1978).

[27] Goodman, J.W., Statistical properties of laser speckle patterns, *Laser Speckle and Related Phenomena*, edited by J.C. Dainty, Springer-Verlag, pp. 9–75 (1975).

[28] Viénot, J.Ch., Froehly, C., Monneret, J. and Pasteur, J., Hologram interferometry: Surface displacement fringe analysis as an approach to the study of mechanical strains and other applications to the determination of anisotropy in transparent objects, *Proceedings of the Symposium The Engineering Uses of Holography, Glasgow, 1968*, edited by E.R. Robertson and J.M. Harvey, Cambridge University Press, pp. 133–150 (1970).

[29] Monneret, J., Exploitation des systèmes d'interférences observables en interférométrie holograph par mesure des déplacements angulaires de l'onde diffractée par l'objet, *Optics Communications*, 2, pp. 159–162 (1970).

[30] Jacquot, P. and Rastogi, P.K., Speckle motions by rigid-body movements in free-space geometry: An explicit investigation and extension to new cases, *Applied Optics*, 18, pp. 2022–2032 (1979).

[31] Jacquot, P., Speckle motions in three-dimensional image fields, *Technical Digest Hologram Interferometry and Speckle Metrology*, edited by Optical Society of America, pp. MA 4.1–4.4 (1980).

[32] Jacquot, P., Photographie de speckles: Exemples tirés de l'analyse de déformation de corps solides, *Lasers et applications industrielles, Cour A VCP 1982*, edited by C. Bonjour and M. Matthey, Presses Polytechniques Romandes, pp. 149–204 (1982).

[33] Rastogi, P.K. and Jacquot, P., Speckle metrology techniques: A parametric examination of the observed fringes, *Optical Engineering*, 21, pp. 411–426 (1982).

[34] Schumann, W. and Dubas, M., Holographic Interferometry, Springer-Verlag (1979).

[35] Archbold, E., Burch, J.M. and Ennos, A.E., Recording of in-phase surface displacement by double exposure speckle photography, *Optical Acta*, 17, pp. 883–898 (1970).

[36] Archbold, E. and Ennos, A.E., Displacement measurement from double exposure laser photographs, *Optica Acta*, 19, pp. 253–271 (1972).

[37] Khetan, R.P. and Chiang, F.P., Strain analysis by one-beam laser speckle interferometry, 1: single aperture method, *Applied Optics*, 15, pp. 2205–2215 (1976).

[38] Ennos, A.E., Speckle interferometry, *Progress in Optics, volume XVI*, edited by E. Wolf, North-Holland Publishing Company, pp. 233–288 (1978).

[39] Speckle Metrology, edited by R.K. Erf, Academic Press (1978).

[40] Coherent Optical Techniques in Experimental Mechanics, *Special issue of Optical Engineering*, 21, pp. 377–495 (1982).

[41] Stetson, K.A., The vulnerability of speckle photography to lens aberrations, *Journal of the Optical Society of America*, 67, pp. 1587–1590 (1977).

[42] Roblin, M.L., Schalow, G. and Chourabi, B., Interférométrie différentielle des aberrations d'un système optique par photographie de speckles, *Journal of Optics*, 8, pp. 149–158 (1977).

[43] Kaufmann, G.H., On the numerical processing of speckle photograph fringes, *Optics and Laser Technology*, 12, pp. 207–209 (1980).

[44] Yamaguchi, I., Speckle displacement and decorrelation in the diffraction and image fields for small object deformation, *Optica Acta*, 28, pp. 1359–1376 (1981).

[45] Tiziani, H.J., Vibration analysis and deformation measurement, *Speckle Metrology*, edited by R.K. Erf, Academic Press, pp. 73–110 (1978).

[46] Ebbeni, J., Application in civil engineering of a sandwich speckle method, *Proceedings of the First European Conference on Optics Applied to Metrology*, SPIE volume 136, pp. 291–293 (1978).

[47] Beranek, W.J. and Bruinsma, A.J.A., Sandwich speckle interferometry, *Proceedings of the 6th International Conference on Experimental Stress Analysis, Munich*, VDI-Verlag GmBH, Düsseldorf, pp. 417–425 (1978).

[48] Tiziani, H.J., Leonhardt, K. and Klenk, J., Real time displacement and tilt analysis by a speckle technique using $Bi_{12}SiO_{20}$ crystals, *Optics Communications*, 34, pp. 327–331 (1980).

[49] Tiziani, H.J., Real-time measurements in optical metrology, *Proceedings of the 5th International Congress Laser 81, Opto electronics in Engineering*, edited by W. Waidelich, pp. 127–138 (1981).

[50] Kaufmann, G.H., Ennos, A.E., Gale, B. and Pugh, D.J., An electro-optical read-out system for analysis of speckle photographs, *Journal of Physics E: Scientific Instruments*, 13, pp. 579–584 (1980).

[51] Ineichen, B., Eglin, P. and Dändliker, R., Hybrid optical and electronic image processing for strain measurements by speckle photography, *Applied Optics*, 19, pp. 2191–2195 (1980).

[52] Smith, G.B. and Stetson, K.A., Heterodyne read-out of specklegram halo interference fringes, *Applied Optics*, 19, pp. 3031–3033 (1980).

[53] Archbold, E., Ennos, A.E. and Virdee, M.S., Speckle photography for strain measurement – A critical assessment, *Proceedings of the First European Conference on Optics Applied to Metrology*, SPIE volume 136, pp. 258–264 (1978).

[54] Pedretti, M.G. and Chiang, F.P., Effect of magnification change in laser speckle photography, *Journal of the Optical Society of America*, 68, pp. 1742–1748 (1978).

[55] Jacquot, P. and Rastogi, P.K., Influence of out-of-plane deformation and its elimination in white light speckle photography, *Optics and Lasers in Engineering*, 2, pp. 33–55 (1981).

[56] Gregory, D.A., Speckle scatter, affine geometry and tilt topology, *Optics Communications*, 20, pp. 1–5 (1977).

[57] Gregory, D.A., Topological speckle and structures inspection, *Speckle Metrology*, edited by R.K. Erf. Academic Press, pp. 183–223 (1978).

[58] Chiang, F.P. and Juang, R.M., Laser speckle interferometry for plate bending problems, *Applied Optics*, 15, pp. 2199–2204 (1976).

[59] Stetson, K.A. and Harrison, I.R., Determination of the principle surface strains on arbitrarily deformed objects via tandem speckle photography, *Proceedings of the 6th International Conference on Experimental Stress Analysis, Munich*, VDI-Verlag, GmBH, Düsseldorf, pp. 149–154 (1978).

[60] Boone, P.M. and De Backer, L.C., Speckle methods using photography and reconstruction in incoherent light, *Optik*, 44, pp. 343–355 (1976).

[61] Burch, J.M. and Forno, C., A high sensitivity moiré grid technique for studying deformation in large objects, *Optical Engineering*, 14, pp. 178–185 (1975).

[62] Chiang, F.P. and Asundi, A., White light speckle method of experimental strain analysis, *Applied Optics*, 18, pp. 409–411 (1979).

[63] Leendertz, J.A., Interferometric displacement measurement on scattering surfaces utilizing speckle effects, *Journal of Physics E: Scientific Instruments*, 3, pp. 214–218 (1970).

[64] Butters, J.N., Electronic speckle pattern interferometry: a general review background to subsequent papers, *Proceedings of the Symposium The Engineering Uses of Coherent Optics, Glasgow 1975*, edited by E.R. Robertson, Cambridge University Press, pp. 155–169 (1976).

[65] Jones, R., Design and application of a speckle pattern interferometer for measurement of total plane strain fields, *Optics and Laser Technology*, 8, pp. 215–219 (1976).

[66] Wykes, C., Use of electronic speckle pattern interferometry (ESPI) in the measurement of static and dynamic surface displacements, *Optical Engineering*, 21, pp. 400–406 (1982).

[67] Jones, R. and Wykes, C., Decorrelation effects in speckle pattern interferometry II, *Optica Acta*, 24, pp. 533–550 (1977).

[68] Spajer, M., Rastogi, P.K. and Monneret, J., In-plane displacement and strain measurement by speckle interferometry and moiré derivation, *Applied Optics*, 20, pp. 3392–3402 (1981).

[69] Butters, J.N. and Leendertz, J.A., Speckle pattern and holographic techniques in engineering metrology, *Optics and Laser Technology*, 3, pp. 26–30 (1971).

[70] Butters, J.N. and Leendertz, J.A., A double exposure technique for speckle pattern interferometry, *Journal of Physics E: Scientific Instruments*, 4, pp. 277–279 (1971).

[71] Duffy, E.D., Measurement of surface displacement normal to the line of sight, *Experimental Mechanics*, 14, pp. 378–384 (1974).

[72] Hung, Y.Y. and Liang, C.Y., Image-shearing camera for direct measurement of surface strains, *Applied Optics*, 18, pp. 1046–1050 (1979).

[73] Hung, Y.Y., Shearography: a new optical method for strain measurement and nondestructive testing, *Optical Engineering*, 21, pp. 391–395 (1982).

[74] Rosenberg, A. and Politch, J., Fringe parameters in speckle shearing interferometry, *Optics Communications*, 26, pp. 301–304 (1978).

[75] Brdicko, J., Olson, M.D. and Hazell, C.R., Theory for surface displacement and strain measurements by laser speckle interferometry, *Optica Acta*, 25, pp. 963–989 (1978).

[76] Hung, Y.Y., Displacement and strain measurement, *Speckle Metrology*, edited by R.K. Erf, Academic Press, pp. 51–71 (1978).

[77] Vest, M.C., Holographic Interferometry, Wiley, New York (1979).

[78] Ostrovsky, Y.I., Butusov, M.M. and Ostrovskaya, G.V., Interferometer by Holography, Springer-Verlag (1980).

[79] Collier, R.J., Burckhardt, C.B. and Lin, L.H., Optical Holography, Academic Press (1971).

[80] Handbook of Optical Holography, edited by H.J. Caulfield, Academic Press (1979).

[81] Welford, W.T., Fringe visibility and localization in hologram interferometry, *Optics Communications*, 1, pp. 123–125 (1969).

[82] Walles, S., On the concept of homologous rays in holographic interferometry of diffusively reflecting surfaces, *Optica Acta*, 17, pp. 899–913 (1970).

[83] Sollid, J.E., Translational displacements versus deformation, displacements in double-exposure holographic interferometry, *Optics Communications*, 2, pp. 282–288 (1970).

[84] Stetson, K.A., Fringe interpretation for hologram interferometry of rigid-body motions and homogeneous deformations, *Journal of the Optical Society of America*, 64, pp. 1–10 (1974).

[85]   Dubas, M. and Schumann, W., Sur la détermination holographie de l'état de déformation à la surface d'un corps non-transparent, *Optica Acta*, 21, pp. 547–562 (1974).

[86]   Leroy, J., Localisation des franges en interférométrie holographique, *Nouvelle Revue d'Optique*, 6, pp. 329–337 (1975).

[87]   Ennos, A.E., Measurement of in-plane surface strain by hologram interferometry, *Journal of Scientific Instruments*, 1, pp. 731–734 (1968).

[88]   Tsujiuchi, J., Takeya, N. and Matsuda, K., Mesure de la déformation d'un objet par interférométrie holographique, *Optica Acta*, 16, pp. 709–722 (1969).

[89]   Dhir, S.K. and Sikora, J.P., An improved method for obtaining the general displacement field from a holographic interferogram, *Experimental Mechanics*, 7, pp. 323–327 (1972).

[90]   Hung, Y.Y., Hu, C.P., Henley, D.R. and Taylor, C.E., Two improved methods of surface-displacement measurements by holographic interferometry, *Optics Communications*, 8, pp. 48–51, (1973).

[91]   King, P.W., Holographic interferometry technique utilizing two plates and relative fringe orders for measuring microdisplacements, *Applied Optics*, 13, pp. 231–233 (1974).

[92]   Dubas, M. and Schumann, W., On direct measurements of strain and rotation in holographic interferometry using the line of complete localization, *Optica Acta*, 22, pp. 807–819 (1975).

[93]   Ebbeni, J. and Charmet, J.C., Strain components obtained from contrast measurement of holographic fringe patterns, *Applied Optics*, 16, pp. 2543–2545 (1977).

[94]   Gates, J.W.C., Holographic measurement of surface distortion in three dimensions, *Optics Technology*, 1, pp. 247–250 (1969).

[95]   Abramson, N., The holo-diagram V: a device for practical interpreting of hologram interference fringes, *Applied Optics*, 11, pp. 1143–1147 (1972).

[96]   Fossati-Bellani, V. and Sona, A., Measurement of three-dimensional displacements by scanning a double-exposure hologram, *Applied Optics*, 13, pp. 1337–1341 (1974).

[97]   Boone, P.M. and De Backer, L.C., Determination of three orthogonal displacement components from one double-exposure hologram, *Optik*, 37, pp. 61–81 (1973).

[98]   Ebbeni, J., Combinaison d'une méthode de moiré et d'une méthode holographique pour déterminer l'état de déformation d'un objet diffusant, *Proceedings of the 5th International Conference on Experimental Stress Analysis, Udine 1974*, CISM, Udine, pp. 4.20–25 (1974).

[99]   Sciammarella, C.A. and Gilbert, J.A., Holographic moiré technique to obtain separate patterns for components of displacement, *Experimental Mechanics*, 16, pp. 215–220 (1976).

[100] Dändliker, R. Eliasson, B., Ineichen, B. and Mottier, F.M., Quantitative determination of bending and torsion through holographic interferometry, *Proceedings of the Symposium The Engineering Uses of Coherent Optics, Glasgow 1975*, edited by E.R. Robertson, Cambridge University Press, pp. 99–117 (1976).

[101] Cadoret, G., Nouvelles méthodes optiques pour l'étude du comportement des structures et matériaux, *Annales de l'Institut Technique du Bâtiment et des Travaux Publics, Série Essais et Mesures*, 373, pp. 28–71 (1979).

[102] Firth, K., Photo induced thermoplastic recording and its applications, *The Marconi Review*, XLI, pp. 61–73 (1978).

[103] Schumann, W. and Dubas, M., On the motion of holographic images caused by movements of the reconstruction light source, with the aim of application to deformation analysis, *Optik*, 46, pp. 377–392 (1976).

[104] Aleksoff, C.C., Temporal modulation techniques, *Holographique Nondestructive Testing*, edited by R.K. Erf, Academic Press, pp. 247–263 (1974).

[105] Sciammarella, C.A. and Chawla, S.K., A lens holographic moiré technique to obtain components of displacements and derivatives, *Experimental Mechanics*, 18, pp. 373–381 (1978).

[106] Gilbert, J.A. and Exner, G.A., Holographic displacement analysis using image plane technique, *Experimental Mechanics*, 18, pp. 382–388 (1978).

[107] Beranek, W.J. and Bruisma, A.J.A., A geometrical approach to holographic interferometry, *Experimental Mechanics*, 20, pp. 289–300 (1980).

[108]   Sciammarella, C.A., Rastogi, P.K., Jacquot, P. and Narayanan, R., Holographic moiré in real time, *Experimental Mechanics*, 22, pp. 52–63 (1982).

[109]   Sciammarella, C.A., Holographic moiré, an optical tool for the determination of displacements, strains, contours and slopes of surfaces, *Optical Engineering*, 21, pp. 447–457 (1982).

[110]   Rastogi, P.K., Spajer, M. and Monneret, J., In-plane deformation measurement using holographic moiré, *Optics and Lasers in Engineering*, 2, pp. 79–103 (1981).

[111]   Dändliker, R. and Eliasson, B., Accuracy of heterodyne holographic strain and stress determination, *Experimental Mechanics*, 19, pp. 93–101 (1979).

[112]   Sciammarella, C.A. and Chawla, S.K., Multiplication of holographic fringes, its application to crack detection, *Proceedings of the 6th International Conference on Experimental Stress Analysis, Munich*, VDI-Verlag GmbH, Düsseldorf, pp. 775–779 (1978).

[113]   Smith, R.W. and Williams, T.H., A depth encoding high resolution holographic microscope, *Optik*, 39, pp. 150–155 (1972).

[114]   Cox, M.E., Holographic microscopy, theory and applications, *Multidisciplinary Microscopy*, SPIE volume 104, pp. 69–72 (1977).

[115]   Rhodes, M.B. and Cournoyer, R.F., Interferometry with a holographic microscope, *Multidisciplinary Microscope*, SPIE volume 104, pp. 21–28 (1977).

[116]   Briones, R.A., Heflinger, L.O. and Wuerker, R.F., Holographic microscopy, *Applied Optics*, 17, pp. 944–950 (1978).

[117]   Cox, M.E. and Vahala, K.J., Image plane holograms for holographic microscopy, *Applied Optics*, 17, pp. 1455–1457 (1978).

[118]   Parks, V.J., The range of speckle metrology, *Experimental Mechanics*, 20, pp. 181–191 (1980).

[119]   Holographic Nondestructive Testing, edited by R.K. Erf, Academic Press (1974).

[120]   Ennos, A.E., Optical holography and coherent light techniques, *Research Techniques in Nondestructive Testing, volume 1*, edited by R.S. Sharpe, Academic Press, pp. 155–180 (1970).

[121]   Sollid, J.E., Optical holographic interferometry, *Research Techniques in Nondestructive Testing, volume 2*, edited by R.S. Sharpe, Academic Press, pp. 185–222 (1973).

[122]   Pflug, L., Application of optical methods to crack detection in bended concrete beams, *Proceedings of the 7th National Congress of the Italian Society for Stress Analysis, supplement*, Cagliari A.I.A.S., pp. 5–26 (1979).

[123]   Evans, W.T. and Luxmoore, A.R., Measurement of in-plane displacements around crack tips by a laser speckle method, *Engineering Fracture Mechanics*, 6, pp. 735–743 (1974).

[124]   Luxmoore, A.R., Measurement of displacements around crack tips, *Speckle Metrology*, edited by R.K. Erf, Academic Press, pp. 257–266 (1978).

[125]   Trafalian, M.A. and Taylor, C.E., Stress wave and propagating crack studies using speckle photography with a pulsed ruby laser, *Proceedings of IUTAM Symp. Optical Methods in Mechanics of Solids, Poitier*, edited by A. Lagarde, Sijthoff and Noordhoff, pp. 337–342 (1981).

[126]   De Backer, L.C., In-plane displacement measurement by speckle interferometry, *Nondestructive Testing*, 8, pp. 177–180 (1975).

[127]   Archbold, A. and Ennos, A.E., Laser photography to measure the deformation of weld cracks under load, *Nondestructive Testing*, 8, 181–184 (1975).

[128]   Chiang, F.P., Application of laser speckles to nondestructive evaluation, *Advances in Laser Engineering*, SPIE volume 122, pp. 2–11 (1977).

[129]   Chiang, F.P. and Asundi, A., A white light speckle method applied to the determination of stress intensity factor and displacement field around a crack tip, *Engineering Fracture Mechanics*, 15, pp. 115–121 (1981).

[130]   Butters, J.N. and Leendertz, J.A., Advances in electro optical techniques and lasers for engineering metrology and NDT, *Proceedings of the Electro-Optics International Conference, Brighton, 1972*, Kiver Communications, pp. 43–57 (1974).

[131]   Butters, J.N., Jones, R. and Wykes, C., Electronic speckle pattern interferometry, *Speckle Metrology*, edited by R.K. Erf, Academic Press, pp. 111–158 (1978).

[132] Barker, D.B. and Fourney, M.E., Three-dimensional speckle interferometric investigation of the stress intensity factor along a crack front, *Experimental Mechanics*, 17, pp. 241–247 (1977).

[133] Dudderar, T.D., Applications of holography to fracture mechanics, *Experimental Mechanics*, 9, pp. 281–285 (1969).

[134] Dudderar, T.D. and O'Regan, R., Measurement of the strain field near a crack tip in polymethylmethacrylate by holographic interferometry, *Experimental Mechanics*, 11, pp. 49–56 (1971).

[135] Vest, C.M., McKague, E.L. and Friesen, A.A., Holographic detection of microcracks, *Journal of Basic Engineering, Transactions of the ASME*, D 93, pp. 237–241 (1971).

[136] Vest, C.M., Crack detection, *Holographic Nondestructive Testing*, edited by R.K. Erf, Academic Press, pp. 289–301 (1974).

[137] Light, M.F. and Luxmoore, A.R., Detection of cracks is concrete by holography, *Magazine of Concrete Research*, 24, pp. 167–172 (1972).

[138] Light, M.F. and Luxmoore, A.R., Crack detection by holography, *Precast Concrete*, 4, pp. 26–28 (1973).

[139] Stroeven, P. and De Haas, H.M., Detection of cracks in concrete by holographic interferometry, *Proceedings RILEM Symposium New Developments in Nondestructive Testing on Non-Metallic Materials, Constantza, Romania*, p. 19 (1974).

[140] De Haas, H.M. and Stroeven, P., Crack detection techniques in concrete materials research, *Stevin Laboratory Report No. 1-75-1* (1975).

[141] Nathan, S.S., Metha, S.D. and Selvaryan, A., Holographic interferometry study of chemical bond of concrete to smooth steel rods, *Journal of Testing and Evaluation*, 6, pp. 284–286 (1978).

[142] Spetzler, H., Scholz, C.H. and Lu, C.P.J., Strain and creep measurements on rocks by holographic interferometry, *Pure and Applied Geophysics*, 112/3, pp. 571–582 (1974).

[143] Holloway, D.C., Application of holographic interferometry to stress wave and crack propagation problems, *Optical Engineering*, 21, pp. 468–473 (1982).

[144] Cadoret, G., Application of holography to the study of structures and materials, *Proceedings of the 1st European Conference on Optics Applied to Metrology, Strasbourg 1977*, SPIE volume 136, pp. 114–126 (1978).

[145] Meyer, L.W., Jüptner, W. and Steffens, H.D., Fracture toughness investigations using holographic, *Laser 75 Optoelectronics Conference Proceedings, Munich*, pp. 203–205 (1975).

[146] Holloway, D.C., Der, V.K. and Fourney, W.L., Dynamic fracture toughness determination from isopachics of a running crack, *Proceedings of the 6th International Conference on Experimental Stress Analysis, Munich*, VDI-Verlag GmbH, pp. 71–76 (1978).

[147] Rossmanith, H.P., Dynamic stress intensity factor determination from isopachics, *Experimental Mechanics*, 19, pp. 281–285 (1979).

[148] Dudderar, T.D. and Doerries, E.M., A study of effective crack length using holographic interferometry, *Experimental Mechanics*, 16, pp. 300–304 (1976).

[149] Krenz, H.G., Kramer, E.J. and Ast, D.G., Direct measurements of the strain energy release rate of propagating crazes by holographic interferometry, *Journal of Polymer Science Polymer Letters Edition*, 13, pp. 583–587 (1975).

[150] Marom, E. and Mueller, R.K., Optical correlation for impending fatigue failure detection, *International Journal of Nondestructive Testing*, 3, pp. 171–187 (1971).

[151] Haworth, W.L., Singh, V.K. and Mueller, R.K., Holographic detection of fatigue-induced surface deformation and crack growth in a high-strength aluminium alloy, *Metallurgical Transactions A*, 11A, pp. 219–229 (1980).

[152] Ohlson, N.G., Experimental determination of crack initiation, *Materials Science III*, 1–2, pp. 15–19 (1977).

Subject index

Angled crack, 6, 43, 55
Acoustic emission, 32, 141
Aggregate
   size, 3, 41
   size and notch-sensitivity, 117
   volume fraction, 3, 7
ASTM, 2, 21, 100, 113

Bond
   cracks, 58, 78, 141
   of aggregate to matrix, 5, 7, 8, 32
   strength, 141

COD criterion, 97, 120
Compliance
   and energy release rate, 118
   definitions, 119
   of double cantilever specimen, 125
   of double torsion specimen, 124
Compressive strength
   and microcracking, 150
   and rate-of-loading, 75
   biaxial, 39
   hysteresis, 36
   multiaxial, 154
   of concrete, 4, 34
   of mortar, 4
   ratio to tensile, 7
   softening, 35
   uniaxial, 34
Concrete
   constituents, 33, 137
   critical energy release rate, 85, 98
   critical strain energy density factor,
      12
   critical stress intensity factor, 85,
      115
   fiber reinforced, 67, 73, 79, 114,
      115, 122
   high-strength, 33, 73, 155
   lightweight, 73
   microcracking, 137, 140
   notch-sensitivity, 77
   polymer impregnated, 67, 73
   R-curves, 132
   surface energy, 81
Cracks
   arrest of, 33
   between dissimilar media, 48
   bridging of, 33
   microcracking, 7, 32
   primary, 5

secondary, 5
   velocity, 95, 96
Creep
   and microcracking, 152
   behavior, 22
   modeling, 23
Critical energy release rate
   G_c, 84, 97
   J_c, 97
   of cement paste, 85, 98
   of concrete, 85, 98
   of mortar, 85, 98
Critical stress intensity factor
   K_c, 111, 113
   K_{Ic}, 111
   K_{Id}, 112
   of cement paste, 85, 115
   of concrete, 85, 115
   of mortar, 85, 115

Damage
   material damage model, 10
   phases, 32
   under time dependent loading, 61
   zone, 16

Elastic modulus
   and rate-of-loading, 75
   and sonic testing, 143
   of concrete, 3, 4
   of mortar, 3, 4
Energy release rate
   and compliance, 118
   G, 2, 79, 118
   J-integral, 80, 97

Fictitious crack model, 80, 97
Fracture resistance
   critical energy release rate, 2,
      79, 84, 97
   critical stress intensity factor,
      2, 84, 115
   fracture toughness, 2, 9, 100
   general, 2

Hardened cement paste
   cracks in, 67
   critical energy release rate, 85, 98
   critical stress intensity factor, 85,
      115
   notch-sensitivity, 77, 117
   surface energy, 81

201